AF839424

Bath University Library

Author HENRY

Bath University Library

28 DAY LOAN

This book should be returned not later than the last date stamped below. The period of loan may be extended on request if no other reader wishes to use the book.

If this book is reserved by another reader it will be recalled and must be returned by the date on the recall notice, or fines will be incurred.

0 8 NOV 1996

17 JAN 1997

8 MAR 2005

2/10/06

1990

1991

UNIVERSITY OF BATH LIBRARY

04 DEC 2025

Withdrawn

ENERGY MANAGEMENT

ENERGY, POWER, AND ENVIRONMENT

A Series of Reference Books and Textbooks

Editor

PHILIP N. POWERS
Professor Emeritus of Nuclear Engineering
Purdue University
West Lafayette, Indiana

Consulting Editor
Energy Management and Conservation
PROFESSOR WILBUR MEIER, JR.
Head, School of Industrial Engineering
Purdue University
West Lafayette, Indiana

1. Heavy Oil Gasification, *edited by Arnold H. Pelofsky*
2. Energy Systems: An Analysis for Engineers and Policy Makers, *edited by James E. Bailey*
3. Nuclear Power, *by James J. Duderstadt*
4. Industrial Energy Conservation, *by Melvin H. Chiogioji*
5. Biomass: Applications, Technology, and Production, *by Nicholas P. Cheremisinoff, Paul N. Cheremisinoff, and Fred Ellerbusch*
6. Solar Energy Technology Handbook: Part A, Engineering Fundamentals; Part B, Applications, Systems Design, and Economics, *edited by William C. Dickenson and Paul N. Cheremisinoff*
7. Solar Voltaic Cells, *by W. D. Johnston, Jr.*
8. Energy Management: Theory and Practice, *by Harold W. Henry, Frederick W. Symonds, Robert A. Bohm, John H. Gibbons, John R. Moore, and William T. Snyder*

Additional Volumes in Preparation

ENERGY MANAGEMENT

THEORY AND PRACTICE

Harold W. Henry   Frederick W. Symonds   Robert A. Bohm
John R. Moore   William T. Snyder
The University of Tennessee
Knoxville, Tennessee

John H. Gibbons
Office of Technology Assessment
U.S. Congress, Washington, D.C.

MARCEL DEKKER, INC.   New York and Basel

UNIVERSITY OF BATH
LIBRARY
12 | 24 MAR 1981
~~620.9~~ 338.26209
51624

Library of Congress Cataloging in Publication Data

Main entry under title:

Energy management.

(Energy, power, and environment ; 8)
Includes bibliographical references and index.
1. Energy policy--United States. 2. Energy conservation--United States. 3. Energy conservation--United States--Case studies. I. Henry, Harold W. II. Series.
HD9502.U52E498 333.79'0973 80-21671
ISBN 0-8247-1094-0

COPYRIGHT © 1980 by MARCEL DEKKER, INC. ALL RIGHTS RESERVED

Neither this book nor any part may be reproduced or transmitted in any form or by any means, electronic or mechanical, including photocopying, microfilming, and recording, or by any information storage and retrieval system, without permission in writing from the publisher.

MARCEL DEKKER, INC.

270 Madison Avenue, New York, New York 10016

Current printing (last digit):
10 9 8 7 6 5 4 3 2 1

PRINTED IN THE UNITED STATES OF AMERICA

PREFACE

This book is designed to provide a broad look at energy resources in the future and ways to manage them more efficiently. It is intended for college and university students, as well as general managers, engineers, and first-level supervisors in any operating facility. The focus is on energy management in one firm or organization, but public policy issues are also outlined and recent responses of business corporations and government agencies are discussed.

Concern about the availability and cost of various types of energy has increased dramatically in the United States, and in all other countries of the world that are not internally energy-sufficient, since the oil embargo by Arab oil-producing nations in 1973 Governments, business firms, other institutions, and consumers no longer can assume that energy of any type in any quantity will be available when and where they want it and at a relatively low price. The demand for energy has increased at a very rapid rate in the past few decades, especially in developed countries like the United States--in part due to government policies, which insured "cheap" energy; in part due to conventional cost accounting methods, which failed to internalize all of the social costs of obtaining and using energy; and in part due to the indifference of users--individuals, businesses, and governments--concerning energy wastefulness.

During the early years of the 1980s, everyone who produces or uses energy or who influences energy production and consumption will be forced to examine current practices and policies and to modify them to cope with a changing world of the future. This will be a world in which fossil fuels will become much more expensive and uneconomical to obtain, in which new sources of energy will be sought and new supply and utilization technologies will be developed, tested, and used; and in which energy consumption patterns will shift markedly due to the economic and political realities of the times.

Thus, this book should help anyone who is concerned about energy to understand the broad problem and specific issues more clearly and to make wiser decisions and take wiser actions in the future.

The material upon which this book is based was developed while the authors were working on a grant from the former Energy Research and Development Administration (ERDA, now part of the U.S. Department of Energy). The grant was for the purpose of developing, teaching, evaluating, and disseminating a course on "energy management."

The authors gratefully acknowledge the support of ERDA, and in particular that of Dr. Melvin H. Chiojiogi, former Assistant Director for Systems Analysis, Division of Buildings and Community Systems in ERDA. Also, the contributions of many students from three classes who have taken the course are appreciated, and a debt of gratitude is expressed to the typist who prepared the manuscript, Susie Kyte.

The opinions expressed are those of the authors, as are any omissions or flaws.

Harold W. Henry
Frederick W. Symonds
Robert A. Bohm
John H. Gibbons
John R. Moore
William T. Snyder

CONTENTS

Part 1

FUNDAMENTAL CONCEPTS OF ENERGY MANAGEMENT

Before an attempt is made to evaluate alternative policies or plans to deal with energy-related issues or problems, several basic concepts of physics, engineering, economics, and management should be understood. Also, a broad historical perspective of energy resource supply and utilization is needed. Part I attempts to provide this fundamental knowledge.

In Chapter 1, a brief overview is presented of the historical usage of wood, petroleum, natural gas, coal, and nuclear fission reactors to provide energy in the United States, as well as of the level of consumption in 1975 for several energy sources in each major consuming sector. Three alternative projections of U.S. energy demand in the year 2000 are given for major energy sources. Also, the total consumption of five types of energy in 13 nations is presented graphically for 1972.

In Chapter 2, physical and engineering concepts of energy, power, and thermodynamics are discussed, as are units for measuring energy and power.

Economic concepts of market, price, supply, demand, elasticity, and equilibrium are reviewed in Chapter 3, and many energy-related illustrations are provided.

Fundamentals of management, organization theory, and decision-making are presented in Chapter 4. The basic functions of managers, organizational goals, the system concept, organization structures, approaches to energy management, strategic planning, and investment decision methods are discussed.

Chapter 1

ENERGY RESOURCES: SUPPLY AND DEMAND

INTRODUCTION

Energy Origins

About 4700 million years ago a supernova occurred. In a matter of probably less than 10 seconds, with neutron densities of thousands of tons per cubic inch, many of our solar system's heavy elements were created in the explosion of that star. About a hundred million years later, our sun and planets condensed by gravitational contraction and now contain the debris of that explosion.*

About 4300 million years later (several hundred million years ago), the carboniferous era began and the planet earth began to store up sunlight in fossil form as oil, gas, coal, and shale. These energy sources were created through photosynthesis in primeval swamps and later buried and preserved through geological evolution. Meanwhile, deep inside the earth, radioactive elements from the supernova's debris released energy and made volcanos and earthquakes, which slowly segregated and concentrated our mineral wealth. Thus, all forms of energy available to us today--nuclear, chemical, electromagnetic, and gravitational--were originally derived from the weakest of all physical forces, gravitation.

*This account of the origin of the universe is the one most widely accepted by cosmologists.

The heavy elements from which we derive nuclear fission energy came from the supernova that occurred just before the solar system condensation, as did those radioactive elements which provide geothermal energy. The energy in coal, gas, and oil is basically sunlight stored in the form of organic molecules. Solar, hydro, and wind forms of energy are derived from the sun today, and wood fuel is created by the sun through the process of photosynthesis. Tides result from conversion of gravitational energy in the earth/moon system, and gravitation also holds together the only steady-state thermonuclear fusion device in the solar system--the sun.

Current Energy Dilemma

Energy production and conversions of all kinds are clearly an integral part of our natural environment. We literally owe it to the stars for providing our rich energy inheritance. Until recent years that inheritance seemed limitless. Each time demand pressures threatened a traditional energy source, new supplies, or even new sources, were brought forth, usually through technological innovation.

We are presently immersed in a complex situation in which proven reserves of gas and oil are being exhausted, coal is becoming more expensive, and world economic power is shifting dramatically due to the distribution of petroleum resources. Three major reactions can be expected:

1. Individuals and organizations will respond to the sharply higher and rising price of energy by controlling their energy expenditures.
2. We will seek new energy resources which can pick up the load as oil and gas are depleted.
3. We will develop public policies to encourage more efficient utilization of current energy supplies (and to increase the supply of readily available energy resources such as coal) while new energy supply technologies are being developed and commercialized.

ENERGY RESOURCES

Oil and Gas

Over the past 125 years the dominant type of energy consumed in the United States has shifted from wood to coal to oil and natural gas (Fig. 1.1). In the mid-1970s, over 75% of the total U.S. energy demand was met by oil (45%) and natural gas (over 30%), as shown in Figure 1.2. Fifty years earlier, these fuels provided only about 20% of total requirements, but in another 50 years competitively priced petroleum and natural gas may be largely exhausted.

The United States is becoming more dependent on external suppliers of petroleum each year (Fig. 1.2). About 35% of our petroleum needs were supplied by other nations in 1972, and this fraction had increased to 42% by 1977. This was due to an increasing total energy demand and reduced availability of domestically produced oil and natural gas (Fig. 1.3).

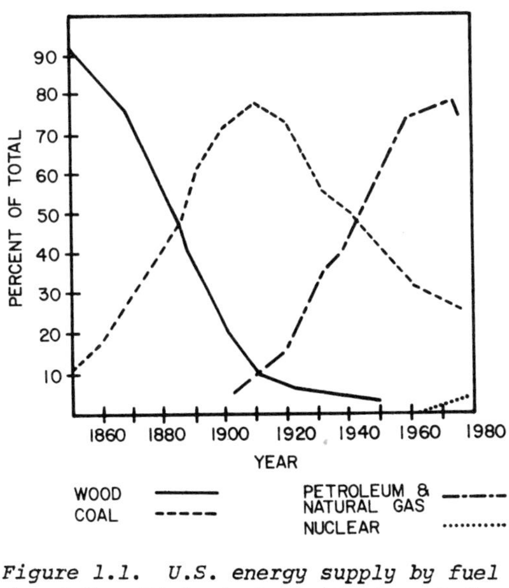

Figure 1.1. U.S. energy supply by fuel type: 1850-1975. (Reproduced from Historical Statistics of the United States Bureau of the Census, *U.S. Bureau of Mines, 1974.)*

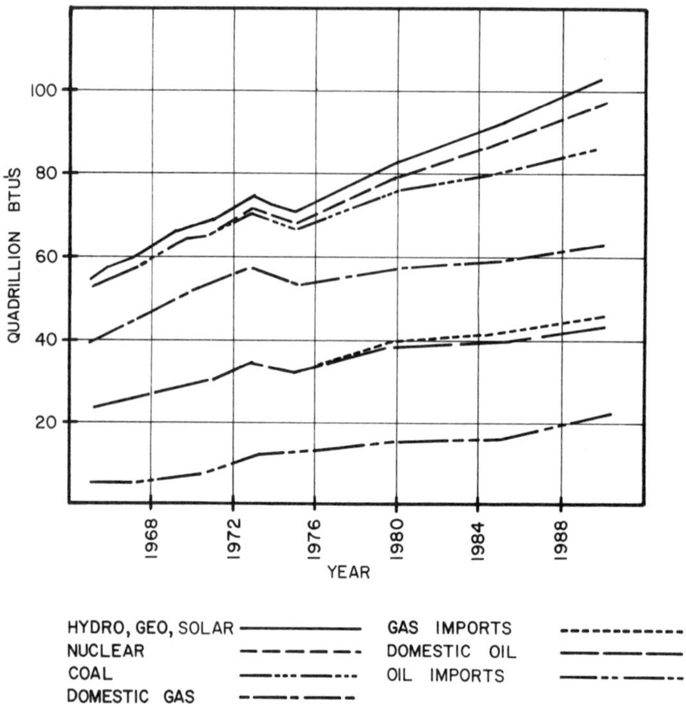

Figure 1.2. U.S. energy supply by fuel type: 1965-1990. (Reproduced from 1977 National Energy Outlook, *Federal Energy Administration, Washington, D.C., 1977 (draft), p. 7.)*

The total production of domestic crude oil, as projected by the U.S. Department of Energy, will increase slowly until the mid-1980s, and then decline by 1990 (Fig. 1.4). Alaskan oil fields will provide an increasing fraction of U.S. oil production and enhanced recovery techniques are expected to boost production for several years. However, conventional production in the contiguous 48 states will decline steadily.

Natural gas production in the United States is expected to decline steadily unless prices are deregulated or other incentives are provided to encourage greater exploration to seek new sources (Fig. 1.5). Other factors which will affect U.S. gas production include geological uncertainties, available investment resources,

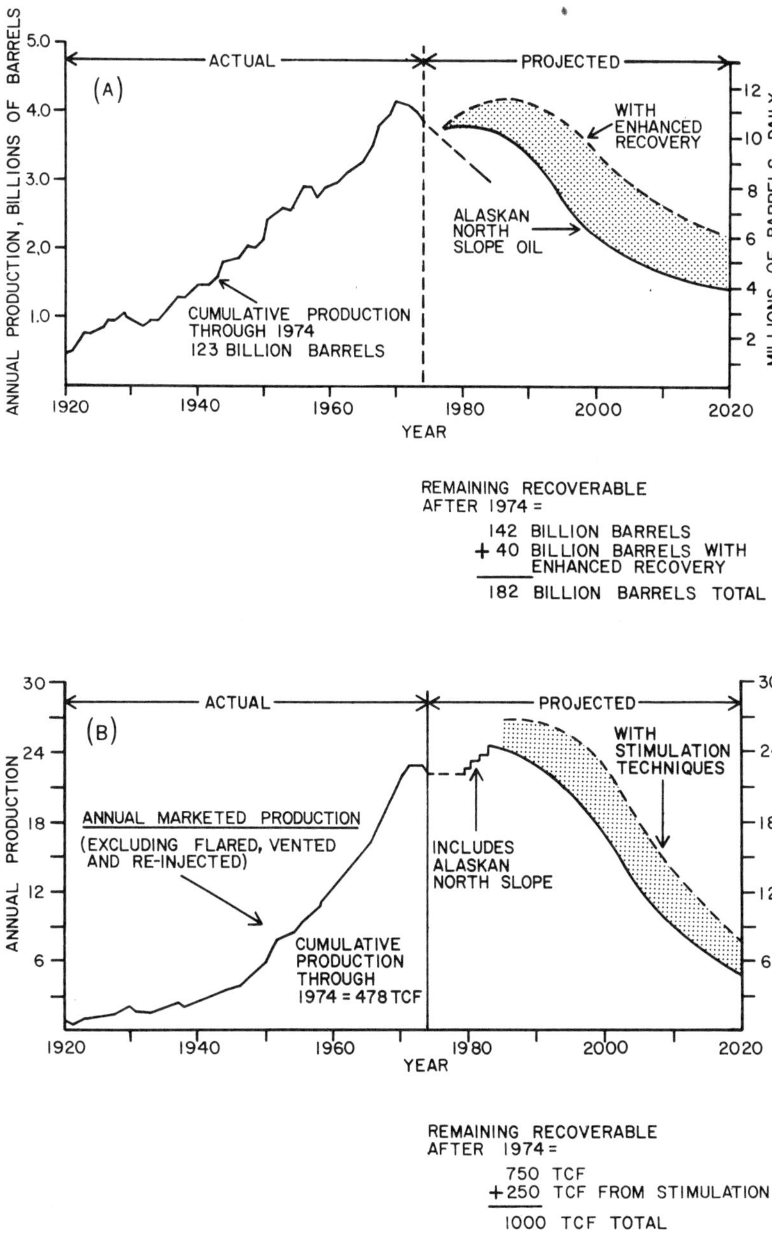

Figure 1.3. (A) Projected U.S. domestic oil production. (B) Projected U.S. domestic natural gas production.

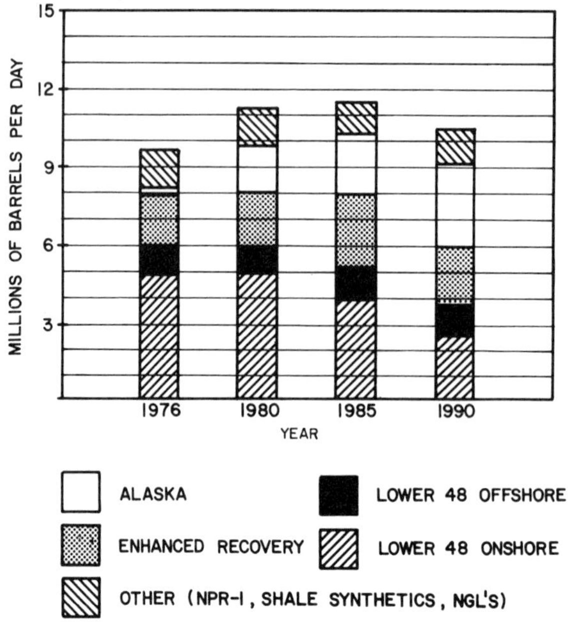

Figure 1.4. Domestic hydrocarbon liquids production: 1976-1990. (Reproduced from 1977 National Energy Outlook, *p. 15.)*

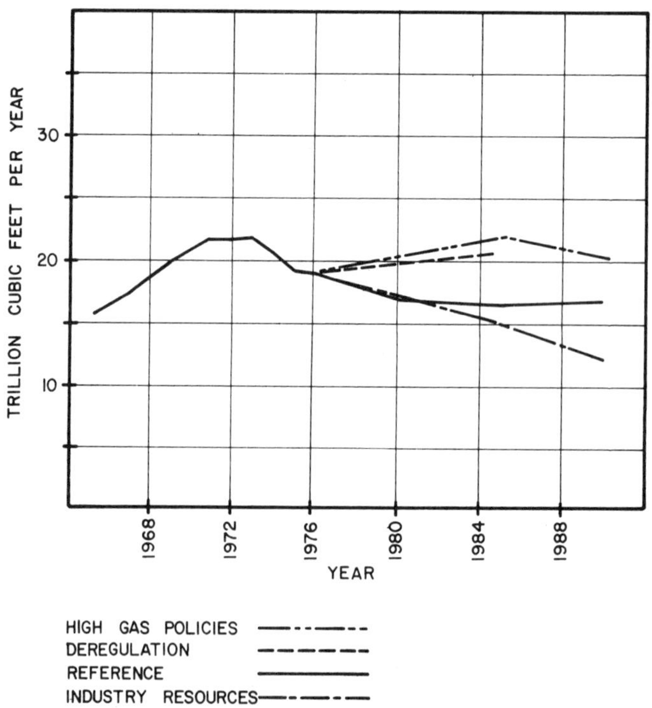

Figure 1.5. U.S. natural gas production: 1965-1990. (Reproduced from 1977 National Energy Outlook, *p. 19.)*

the supply and prices of other sources of energy, and other government policies to encourage gas production.

Even the optimistic experts do not project the supply of domestic crude oil and gas to be large enough to meet our growing demand for energy in future years. Eventually, this same condition will exist in all nations of the world which produce oil and gas. Thus, the dominant position of petroleum and natural gas in the world energy supply will erode slowly over the next several decades.

Coal

The amount of each major type of energy consumed in the United States during the early 1970s is shown in Figure 1.6, as well as the projected consumption to the year 1990 in a base case. In this projection, the consumption of coal is expected to double between

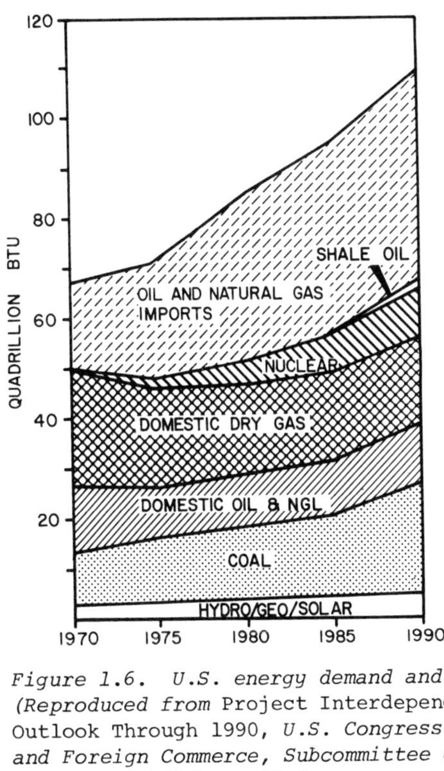

Figure 1.6. U.S. energy demand and supply projections: 1970-1990. (Reproduced from Project Interdependence, U.S. and World Energy Outlook Through 1990, *U.S. Congress, House Committee on Interstate and Foreign Commerce, Subcommittee on Energy and Power, Washington, D.C., Dec. 1977, p. 57.)*

1970 and 1990. Another projection of the quantity of coal to be mined in 1985, compared with production in 1965 and 1975, is shown in Figure 1.7. This graph also indicates the source of the coal produced--eastern United States versus western United States and deep versus surface mining--for the same three years. The major increases in 1985 are expected to come from eastern deep mining and western surface mining.

Coal is the most abundant fossil fuel in the United States, with proven reserves large enough to meet *current levels* of demand for several hundred years [1]. Also, one or more known methods of converting coal to gas and liquids may become economically attractive within a few years, due to the expected higher prices of oil and natural gas. In addition, technological advances for controlling pollutants emitted from coal-fired facilities should make coal more attractive as a fuel.

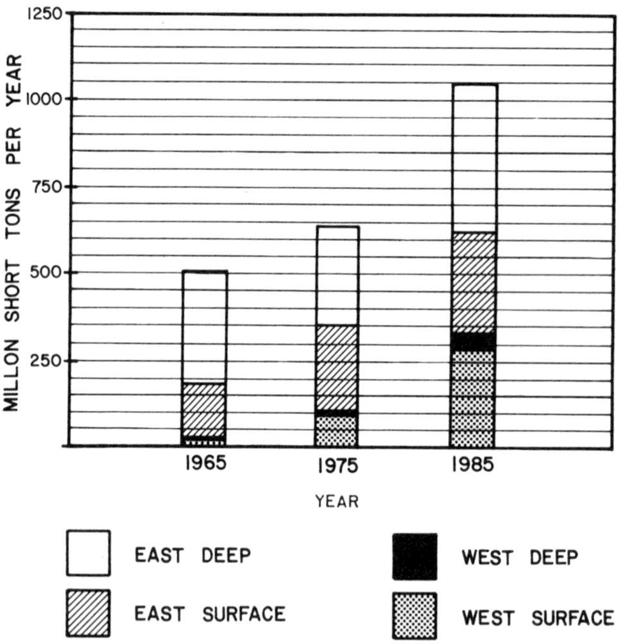

Figure 1.7. U.S. coal production: 1965-1985. (Reproduced from 1977 National Energy Outlook, *p. 23.)*

On the other hand, pollution control devices are expensive, as are land reclamation projects at coal-mining sites. The large investment in equipment to mine and transport coal safely, uncertainties in future energy prices and pollution control regulations, the requirements for skilled manpower, and interruptions such as labor union strikes constitute some of the constraints on the rapid expansion of coal supply and utilization.

Nuclear Fission

The number of nuclear-powered electrical generating plants has increased steadily in the United States during the past two decades, and as of 1977 approximately 70 light water reactors were generating 2.7 quads* of electrical energy. Even so, this output was only 3.6% of the total U.S. energy supply in 1977 [2], and the rate of growth of nuclear power plants has been much slower than was projected in the 1960s. Construction delays, higher and higher costs, burdensome regulatory requirements, environmental concerns, and strong public resistance in some areas have all been contributing factors.

Light water (nonbreeder) nuclear plants have a limited future, since reserves of naturally occurring fissionable material may be exhausted in a half century. However, if breeder reactors (which produce new fuel) were constructed rather than nonbreeder types, these reserves would last many centuries. But considerable public resistance to the development of breeder technology emerged during the 1970s, so the rate of additional development of this source of energy is uncertain.

In spite of many delays and obstacles in the development of nuclear power, most analysts consider nuclear and coal the two energy sources that must be counted on to meet growing U.S. electrical energy needs during the next several decades.

---

*One quad is a quadrillion (10^{15}) Btu (British thermal units).

Other Energy Resources

Solar energy is economical today in many regions for a few applications such as water and space heating, but much development and/or higher prices for conventional sources will be required before it is competitive for most other uses. The contribution of solar energy in meeting total U.S. energy needs is not expected to be significant for many years due to the relatively high costs of applying current solar technology, the large investments required, and the time lags involved in converting from existing to new systems.

Other sources of energy which are being developed include geothermal, nuclear fusion, wind, and tides. These types will be discussed in more detail in Chapter 10.

ENERGY UTILIZATION

Worldwide Energy Demand, by Nation

Consumption of energy by major type and country in 1972 is shown in Table 1.1. The United States uses about 33% of the world's energy supply, even though its population is only 6% of the world's total.

Table 1.1. Energy Consumption by Type, Selected Nations in 1977

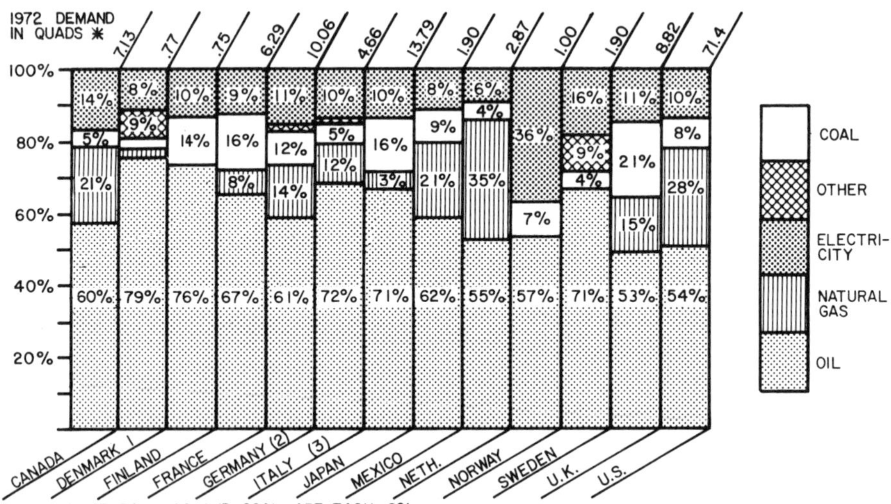

1 NATURAL GAS AND COAL ARE EACH 2%
2 OTHER IS 2%
3 OTHER IS 1%

✱ 1 QUAD (10^{15} BTU) = 1055×10^{15} JOULES

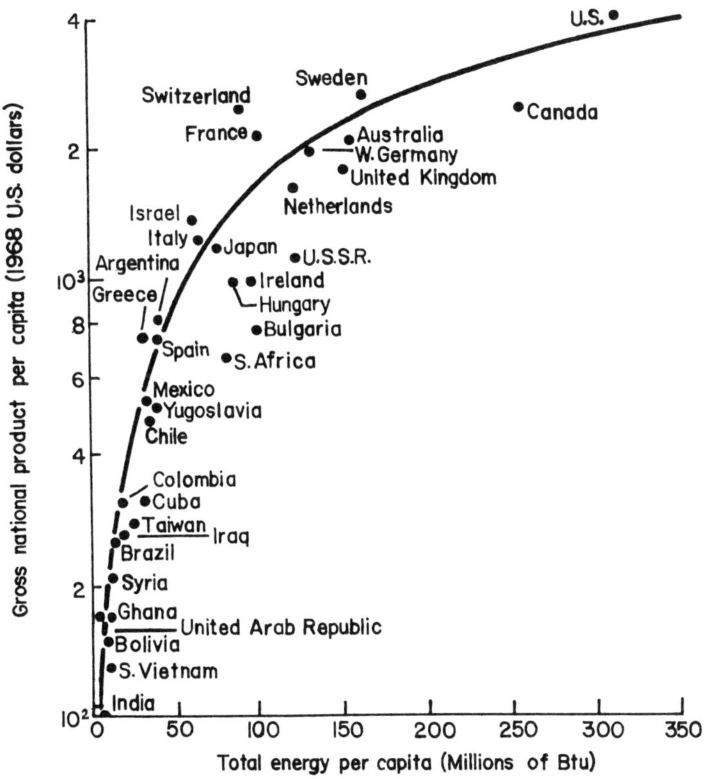

Figure 1.8. Energy utilization intensity by nations in 1968. (Reprinted from Ref. 1, p. 5, Fig. 1.4, with permission.)

During 1950-1970, energy demand grew at an annual rate of 5% worldwide, but since 1973 this growth rate has dropped sharply due to the effects of higher price and efforts to reduce demand.

The intensiveness of energy utilization varies quite widely among nations of the world. It is very high in the United States, which used 337 million Btu per capita in 1971, whereas India used approximately 10 million Btu per capita [3]. Energy utilization per capita is related to gross national product (GNP) per capita for several nations in the world in Figure 1.8. These data indicate that the United States consumes the greatest amount of energy per unit of GNP of any nation in the world; however, Luxembourg is actually a more intense energy user, due to its concentration of steel mills, with a usage of approximately 550 million Btu per capita in the early 1970s.

U.S. Energy Consumption Patterns

Energy consumption in the United States grew at an annual rate of 3.6% during the 20-year period preceding the 1973 oil embargo, and reached the level of about 73 quads in 1975 [4]. The historical consumption by major sectors from 1950 to 1974 and the projected demand to 1990 are shown in Figure 1.9. Two forecasts for future demand are shown, one made before the 1973 oil embargo by the Arab oil-producing nations and one made in 1977 [5]. The more recent projections are significantly lower, due to the expected effects of higher prices and energy conservation efforts.

Another method of reporting energy consumption is to separate electrical energy generated and used by all sectors from the consumption of other types of energy by each sector. A summary of U.S.

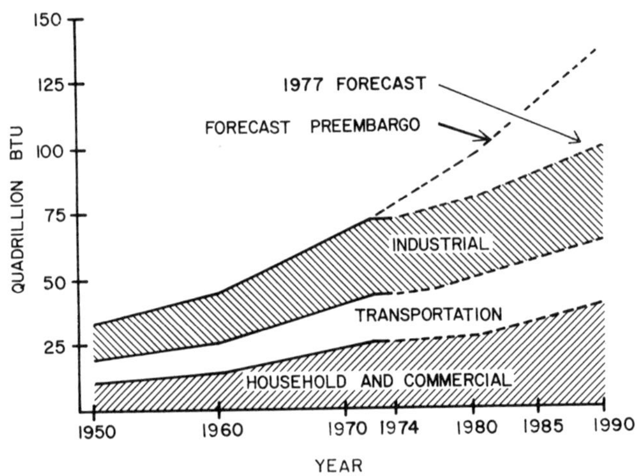

Figure 1.9. U.S. energy consumption by sector: 1950-1990. (Reproduced from Ref. 4, p. xxiv. The 1977 forecasts were revised downward to reflect data developed by the Committee on Nuclear and Alternative Energy Systems. See Science 200, *Vol. 142, 1978.)*

energy consumption in 1975 is shown in Table 1.2 for each type of energy used. These data indicate that one-half of petroleum usage was for transportation, while coal was used primarily for generation of electricity. Natural gas was used very heavily in the industrial sector as well as in the household and commercial sectors.

The flow pattern of energy by source and using sector is illustrated in Figure 1.10. This diagram includes a new dimension, the amount of the total energy consumed which is used productively and the amount which is rejected or wasted. Since approximately 50% of total energy consumption is not used beneficially, there is great potential for more efficient energy conversion and utilization.

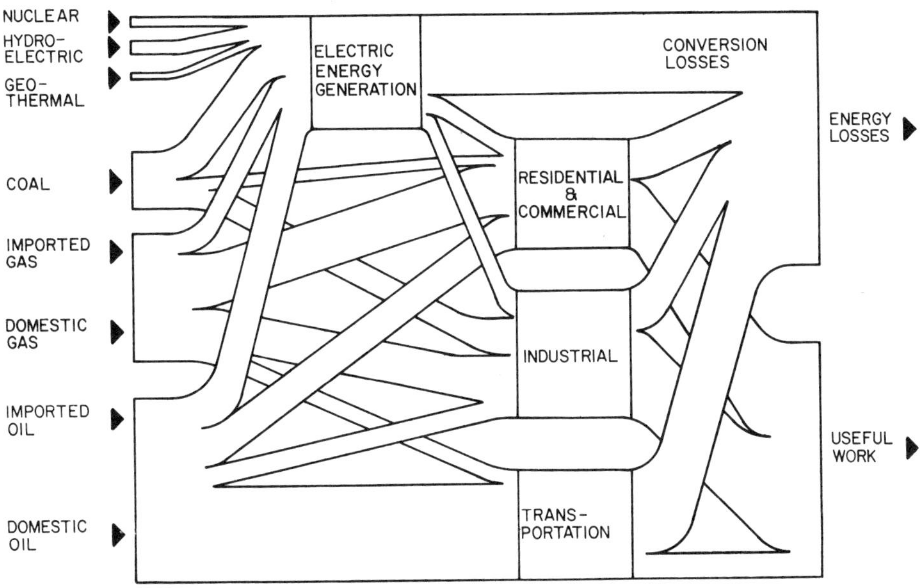

Figure 1.10. U.S. energy flow patterns, 1973. (Reproduced from The National Energy Dilemma, *Joint Committee on Atomic Energy, 1973.)*

Table 1.2. 1975 U.S. Energy Consumption by Major Sectors[a] (Trillion BTU)

	Coal	Petroleum and liquefied natural gas[b]	Natural gas	Hydro power	Nuclear	Total	Percent
Household and commercial	246	5,752	7,589	-	-	13,587	19.3
Industrial	3,821	5,517	8,551	35	-	17,924	25.4
Transportation	1	17,933	595	-	-	18,529	26.2
Electricity generation by utilities	8,760	3,239	3,213	3,194	1,833	20,239	28.7
Miscellaneous and losses	-	278	-	-	-	278	0.4
Total	12,828	32,719	19,948	3,229	1,833	70,557	100.0

[a]Estimated.

[b]Includes natural gas liquids, liquefied refinery gas, and still gas.

Source: Division of Interfuels Studies, Office of Assistant Director--Fuels, Bureau of Mines, U.S. Department of Interior.

SCENARIOS OF THE FUTURE

Scenario analysis is a modern technique for considering complex, interrelated systems under conditions of uncertainty. The technique has been widely used for analysis of potential future energy supply and demand in the United States. Numerous scenarios have been created, with each presenting a possible picture of the future under a particular set of assumptions. The principal value of the methodology is its potential for gaging the effect of various assumed conditions on critical performance variables such as the level of oil imports. Also, scenario analysis is useful for evaluating alternative public policies as well as alternative business strategies under various assumed conditions.

Key Parameters

The construction of scenarios of future U.S. energy supply and demand requires assumptions about such factors as energy prices, technological change, and policy alternatives. In this section, we focus on two factors that will have a critical impact on future U.S. energy markets: (1) the price of imported oil and (2) the rate of economic growth, as measured by the increase in gross national product.

Three Alternative Scenarios

Three scenarios of U.S. sources and uses of energy for the year 2000 are presented in Table 1.3. These scenarios represent high, medium, and low energy growth situations. In column 1, the 1975 situation is shown as a reference point.

The "high" scenario is the historical growth scenario presented in the Ford Foundation Energy Policy Project's Final Report, *A Time to Choose*. It was selected to represent historical growth or, to use the term popularized in Project Independence, "business as usual." This scenario shows where our energy-intensive economic system could have taken us if the trends of the sixties and early seventies had continued. By the year 2000, energy use, according

Table 1.3. Alternative Energy Growth Scenarios for the Year 2000

	1975	High growth[a]	Medium growth[b]	Low growth[c]
1. Total energy				
Energy use (quads)	70.6	187.4	138.8	101.4
Rate of growth (%)	-	3.2	2.4	1.4
2. Energy supplies (quads, %)				
Domestic oil and gas	38.7 (55.0)	83.4 (44.4)	40.1 (28.9)	37.8 (37.2)
Imported oil and gas	13.9 (19.5)	9.0 (4.8)	20.5 (14.8)	7.8 (7.7)
Total oil and gas	52.6 (74.5)	92.3 (49.2)	60.7 (43.7)	45.6 (44.9)
Coal	12.8 (18.1)	44.1 (23.5)	35.3 (25.4)	31.5 (31.1)
Nuclear	1.8 (2.5)	40.2 (21.4)	28.5 (20.5)	14.7 (14.5)
Other	3.0 (4.8)	16.9 (9.0)	15.0 (10.8)	7.6 (7.8)
3. Assumptions				
GNP growth (%), 1975 to 2000	-	3.2	3.1	3.0
Oil price increase (%), 1975 to 2000	-	4.3	3.7	3.5

[a] *A Time to Choose: America's Energy Future*, Energy Policy Project of the Ford Foundation, Baldinger Press, Cambridge, Mass., 1974.

[b] U.S. Energy Research and Development Administration, *Analysis of Alternative Energy Futures*, Washington, D.C., Feb. 17, 1977.

[c] *Implications of a Nuclear Moratorium*, Institute for Energy Analysis, Oak Ridge, Tenn., 1976.

to this scenario, would be 187.4 quads, or more than double the 1975 consumption.

In contrast to the high scenario, the "medium" and "low" scenarios project energy use for the year 2000 of 138.8 quads and 101.4 quads, respectively. In comparing these scenarios, we see that the energy use growth rates are 3.2% per year in the high scenario, 2.4% per year in the medium scenario, and 1.4% per year in the low scenario.

The composition of energy use by source under each scenario is shown in item 2 of Table 1.3. With high growth, total oil and gas consumption retains its relative importance through the year 2000, with total use standing at 92.3 quads. This scenario projects only a small degree of oil imports and assumes a rapid increase in oil prices, as shown in item 3 in Table 1.3. Apparently, this scenario reflects the optimism about oil and gas resources that prevailed in the days preceding the "energy crisis," as the impact of rapid oil price changes on consumption was not significant. The other two scenarios reflect the more realistic views that have emerged in the years since 1973. Under the medium energy growth scenario, for example, rapid oil price growth decreases total oil and gas consumption, but imports remain high. In the low energy growth scenario, high oil price growth lowers both total consumption and the level of imports of oil and gas.

Perhaps the most important relationship found in Table 1.3 is the correlation between energy growth and economic growth. As the energy usage growth rate falls from the high to the low scenario, so does the assumed rate of increase in GNP. Unfortunately, this correlation is often subjected to spurious interpretation. To some analysts, it implies that economic growth causes increased energy consumption, and that to reduce the rate of growth of energy consumption, it is necessary to lower the rate of economic growth. Such an interpretation is both naive and dangerous, especially for time periods of several years or more, because it fails to recognize that the efficiency of energy utilization is an important variable.

Increased efficiency requires better energy management, but it does not necessarily require an economic slowdown.

Future Role of Energy Management

As low cost energy resources became scarcer and supplies more uncertain in the mid-1970s, the price of energy increased significantly. Since the long-term potential for more efficient energy utilization is great, the factors of higher price, less secure supplies, and greater dependence on imports make energy conservation desirable and energy management an economic imperative. Sound energy management will enable us to provide goods and services with less energy required per unit and will accelerate, or at least maintain, the long-term downward trend of energy consumed per unit of GNP.

The major purpose of this book is to provide a procedural framework for planning, implementing, and controlling an energy management program in factories, offices, homes, and other institutions. The procedures outlined can be followed by anyone with responsibility for management, planning, operations, engineering, or maintenance. More specialized technical assistance and reference materials may be required for some energy conservation efforts, but once a formal energy management program is established, these needs will be obvious and easily filled.

REFERENCES

1. Richard C. Dorf, *Energy, Resources, and Policy,* Addison-Wesley, Reading, Mass., 1978, p. 95.
2. *Monthly Energy Review,* Energy Information Administration, U.S. Department of Energy, Oct. 1978.
3. Dorf, *op. cit.*, pp. 5, 7.
4. Federal Energy Administration, *1976 National Energy Outlook*, U.S. Government Printing Office, Washington, D.C., 1976, p. xxiv.
5. Demand and Conservation Panel of the Committee on Nuclear and Alternative Energy Systems, "U.S. Energy Demand: Some Low Energy Futures," *Science 200,* April 14, 1978, pp. 142-152.

Chapter 2

PHYSICAL ASPECTS OF ENERGY

INTRODUCTION

Management under conditions of limited energy availability requires a variety of skills. The manager may not always have an engineering background, but still must be familiar with engineering principles and practices in order to fully understand the options that are available. The engineering aspects of energy are broad in scope and application. The purpose of this chapter is to introduce some of the basic engineering principles which are applicable in making energy decisions. The chapter covers the concept of energy and the related concept of power. The laws of thermodynamics which are applicable to energy transformations of a physical process are discussed and explained through the use of commonly occurring examples.

THE CONCEPT OF ENERGY

Energy is the ability to produce an effect through the transfer of heat or work. Thermodynamics recognizes that the transfer of energy to or from a system can be accomplished only through a work effect or a heat effect. Work transfer occurs if a force moves through a distance, and heat transfer is the result of energy flow due to a temperature difference.

Forms of Energy

Natural gas contains energy since it has the ability to produce heat by combustion. The wind has energy since it has the ability, through the medium of some form of windmill, to exert a force through a distance and thus produce work. Water at an elevation higher than a reference elevation possesses energy by virtue of elevation which is termed potential energy. The energy possessed by the motion of the wind is called kinetic energy. Energy contained in fuels is referred to as chemical energy. Thus, energy may occur in many different forms. Some of the forms of energy which are important from an engineering point of view and the major transformations of energy from one form to another which may be accomplished with current technology are shown in Figure 2.1. Potential energy and kinetic energy are included in the box marked mechanical energy. The arrows represent possible transformations

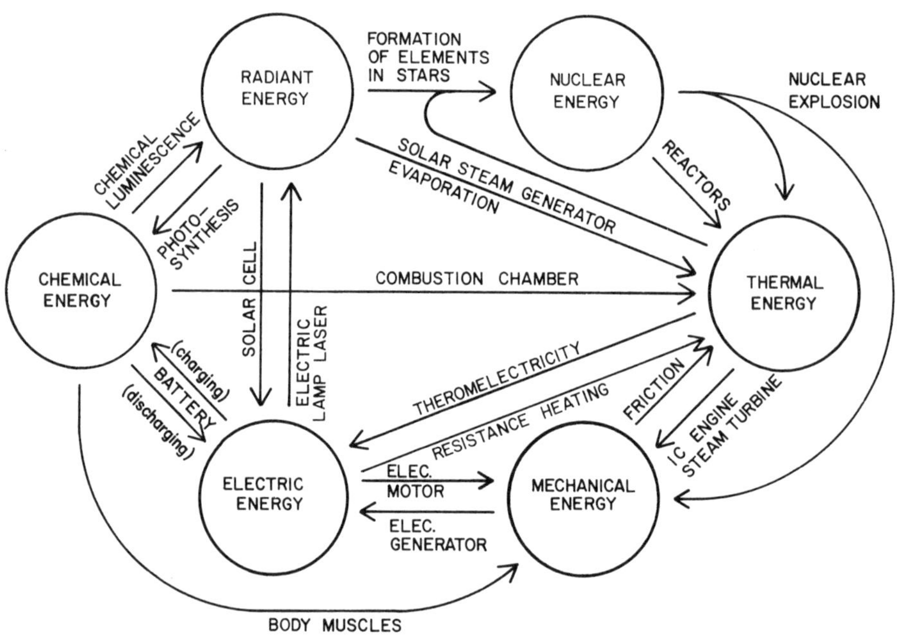

Figure 2.1. Interconnecting energy forms. (Reprinted from Ref. 1, p. 22, with permission.)

of energy among energy forms on the basis of current technology. For example, there is no arrow from nuclear energy to electrical energy because the only technology available for producing electricity from nuclear energy requires intermediate transformations to thermal and mechanical energy.

The Interchangeability of Energy Forms

Heat and work are manifestations of energy transfer in a process. As energy in the form of heat is added to a body, its temperature may rise, or it may remain constant if a change of phase is occurring, such as the conversion of liquid water to a vapor. In a system in which material flows through an industrial plant due to gravity, work is being done by the gravitational force acting on the moving material. Work is done due to a change in the potential energy possessed by the material at the beginning of the process. The terms heat and work are sometimes incorrectly used synonymously with energy content. Heat and work refer only to energy transfer occurring in a process. It is incorrect to speak only of the heat or only of the work possessed by a body; it is an experimentally observable fact that heat and work are always interchangeable to some degree. It is possible to completely convert work into heat although it is never possible to completely convert heat into work.

Our energy resources consist of materials in which energy is stored in a number of different forms. When this stored energy can be conveniently converted to other forms of energy, the materials are referred to as fuels. Fuels are, therefore, materials in which energy is stored in one form and from which energy is extracted and converted to another form. In the example of natural gas as a fuel cited earlier, the energy form in the natural gas is chemical energy which may be converted to thermal energy or heat through the process of combustion.

The most commonly used fuels are the fossil fuels. These fuels store energy in the chemical form. The energy is released in the process of combustion and appears initially in the form of heat.

The heat is used in a variety of ways to achieve desired goals. The two most common ways in which this heat is utilized are in direct heating, such as space heating or industrial processes where parts are heated for annealing, melting, drying, or other purposes, and in its partial conversion to mechanical energy through steam and gas turbines and internal combustion engines. In both internal and external combustion engines the production of work depends on the expansion of a fluid as it is subjected to the heat derived from the combustion of the fuel. In the automobile engine the air and combustion gases in the cylinder expand, as a result of the heat generated by combustion of the fuel, to drive the piston, thus doing mechanical work. In a boiler and steam turbine system the heat from combustion of the fuel converts water into high pressure steam which is allowed to expand through the turbine, again doing mechanical work. In both examples the mechanical work is used directly, to propel automobiles or ships, for example, or indirectly, by conversion to yet another form of energy such as electricity.

These familiar examples illustrate how energy undergoes a number of transformations, utilizing our current technologies to take advantage of the interchangeability of energy forms. Unfortunately, it is not generally possible to obtain these transformations without some losses. These limitations on the interchangeability of energy forms are governed by the first and second laws of thermodynamics, which will be briefly described later in this chapter. The technologies for interchanging energy forms have become an essential part of our energy utilization processes.

The Measurement of Energy

Certain physical laws describe the conversion of energy from one form to another form. The laws governing energy transformations are discussed later in this chapter but first it is necessary to discuss the measurement of energy and to become familiar with the units in which energy is measured. There are a number of systems of units and it would serve the reader well to differentiate

Table 2.1. Conversion Factors

1 British thermal unit (Btu)	Heat required to raise the temperature of 1 lb of water by 1°F
1 kilowatt-hour (kWh)[a]	3,412 Btu
1 barrel crude oil[b]	6,500,000 Btu
1 therm	100,000 Btu
1 kilowatt (kW)	1.341 hp
1 horsepower (hp)	0.746 kW
1 kilowatt-hour (kWh)	1.341 hp-hr
1 horsepower-hour	0.746 kWh
1 ton of refrigeration[c]	12,000 Btu/hr
1 degree day	65F minus mean temperature of the day, °F

[a]To generate 1 kWh requires 10,000 Btu of fuel burned by an average utility.

[b]One U.S. barrel equals 42 U.S. gallons.

[c]One ton of refrigeration requires about 1 kW (or 1.341 hp) in commercial air conditioning.

clearly between them. Essentially, energy units developed through two channels--the English system and the centimeter-gram-second (cgs) system, which later was modified to the meter-kilogram-second (mks) system. This system, in turn, formed the basis for the System Internationale (SI) units. The use of all these systems lingers on in different branches of science and engineering, thus causing considerable confusion for the layperson. This chapter uses units from various systems as needed to reflect the use in the associated industry or branch of engineering.

The amount of heat energy required to raise the temperature of one pound of water by 1°F is defined as a British thermal unit (Btu). This unit is one of the most commonly used units today in the United States. Some useful conversion factors and relationships are listed in Table 2.1. In SI units the standard unit of energy is the joule, which is an inconveniently small amount of energy for engineering purposes. The megajoule, which is one million joules, is sometimes used. Other units of energy are the kilowatt-hour, a common measure of electrical energy which is equivalent to 3,412 Btu, and the therm, a measure of energy frequently used by the suppliers of natural gas, which is equivalent to 100,000 Btu. The energy content of delivered natural gas varies somewhat with the source of the gas, but for most sources the heat energy released in burning one cubic foot is approximately 1,000 Btu. Gas is frequently metered in hundreds of cubic feet (or CCF) so that the energy content of 1 CCF of gas is approximately 100,000 Btu, or 1 therm. The heating value of some of the commonly used fuels is shown in Table 2.2. The values shown represent the heat energy that would be released through combustion of the fuel with oxygen. Such a table of Btu content permits the calculation of the equivalent energy content of different fuels. An energy unit commonly used by many governmental agencies is the barrel of oil equivalent, which is frequently employed to

Table 2.2. BTU Content of Fuels

Energy Source	BTU equivalent
Electricity	3,412 Btu/kWh
Coal	
Anthracite	13,900 Btu/lb
Bituminous	14,000 Btu/lb
Subbituminous	12,600 Btu/lb
Lignite	11,000 Btu/lb
Fuel oil and middle distillates	
Kerosene	134,000 Btu/gal
No. 2 burner fuel oil	140,000 Btu/gal
No. 4 heavy fuel oil	144,000 Btu/gal
No. 5 heavy fuel oil	150,000 Btu/gal
No. 6 heavy fuel oil, 2.7% sulfur	152,000 Btu/gal
No. 6 heavy fuel oil, 0.3% sulfur	143,800 Btu/gal
Gas	
Natural	1,000 Btu/ft^3
Liquefied butane	103,000 Btu/gal
Liquefied propane	91,600 Btu/gal

Source: *Energy Conservation Program Guide for Industry and Commerce*, NBS Handbook 115 (Washington, D.C.: U.S. Government Printing Office, 1974).

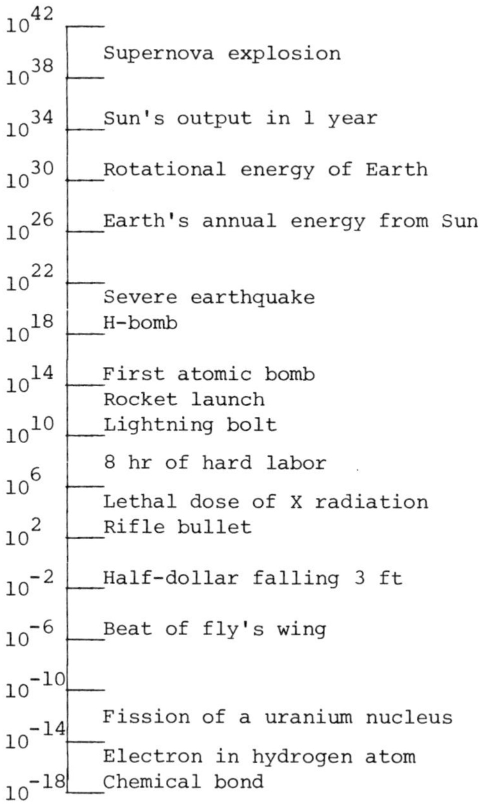

Figure 2.2. Range of energies in physical processes (joules). (Reprinted from Ref. 1, p. 22, with permission.)

express the energy equivalent of various forms of energy. For example, one ton of bituminous coal (14,000 Btu/lb) contains 28 million Btu.* If we assume the energy content of crude oil to be 155,000 Btu/gallon, then a barrel (42 gallons) of crude contains 6.5 million Btu and one ton of coal is seen to be equivalent in energy content to 4.3 barrels of crude oil.

The range of energy levels involved in naturally occurring physical processes is very wide. This fact is illustrated in Figure 2.2. In comparison with the highest energy levels of physical

---

*The abbreviation MBtu is frequently used for million Btu. Unfortunately, the symbol M does not universally indicate the multiple of a million; in some cases it indicates thousands. For this reason, MBtu should be used with caution.

processes, the human effort to control energy transformations covers a relatively small part of the energy spectrum.

THERMODYNAMIC LIMITATIONS

The transfer of energy from one body to another and the transformation of energy from one form to another are governed by the basic laws of thermodynamics.

The First Law of Thermodynamics

The first law states that energy can be neither created nor destroyed. It can only be transformed from one form to another. The concept of energy transformation has already been discussed. This law expands our understanding of the transformation by requiring that no energy be lost in the process. Thus, energy is always conserved, and so the thermodynamicist familiar with the first law might be inclined to question the semantics of the term energy conservation as it is widely used today. A more precise use of thermodynamics terminology would make a distinction between available energy and unavailable energy. A complete treatment of the concept of available and unavailable energy is beyond the scope of this textbook. The qualitative point to be made is that even though energy is always conserved in a process, only a portion of the energy can be converted to work, and this portion is called the available energy.

An example of the first law is the process of converting electrical energy to light (radiant energy). Only 2% of the electrical input to an incandescent light bulb is converted to useful light, which is the desired end use of the energy. The first law tells us that the remaining 98% has not disappeared. It has been converted largely to heat in the form of radiation not visible to the human eye.

As another example, the conversion of the energy in coal to electrical energy in a central power plant involves a sequence of transformations, illustrated in schematic form in Figure 2.3.

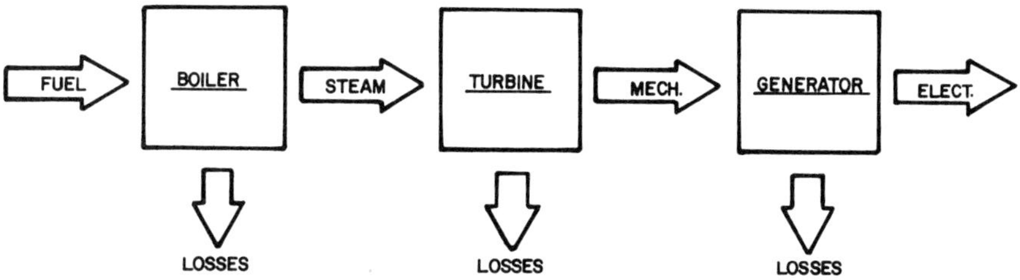

Figure 2.3. Electric power plant.

Chemical energy in the coal is released as heat in the combustion process. A large fraction of this heat is transferred to water in the boiler to produce steam, but some of the heat is lost up the stack in the hot combustion gases or as losses through the walls of the boiler. Nevertheless, all the chemical energy initially in the coal must be accounted for. The thermal energy which is in the steam is partially transformed to mechanical energy in the turbine, but a portion of the energy in the steam cannot be converted to mechanical energy due to a fundamental limitation imposed by the second law of thermodynamics, which will be discussed in the next section. Much of the energy remaining in the steam after it leaves the turbine is transferred to the condenser water so that the steam may be returned to the liquid state. Again, all the energy in the steam can be accounted for although only part of it is available to drive the electric generator for the final transformation to electrical energy. As in the other transformations, some energy is lost and appears mainly as heat in the generator but large electric generators themselves are very efficient. As much as 99% of the input mechanical energy may be available as output electrical energy. The electrical energy output is about 35% of the chemical energy contained in the coal fed to the boiler. The remainder of the energy has not been lost, due to the first law limitation, but appears in forms other than the desired end form of energy--namely, electrical energy. A large part of the difference between the chemical energy contained in the coal and the electrical energy produced by the generator appears as heat energy transferred to the atmosphere or a large body of cooling water such as a river, ocean, or lake.

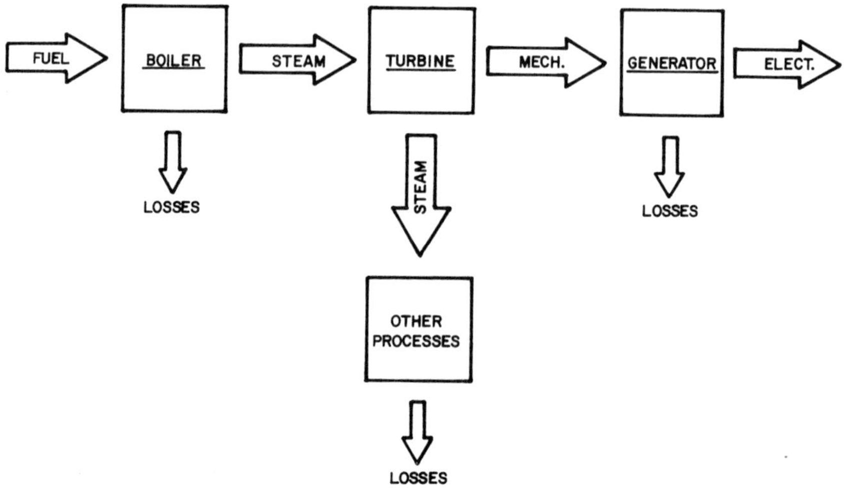

Figure 2.4. Cogeneration of process steam and electricity.

In recent years, as a result of increased awareness of the importance of energy management, considerable attention has been given to the possibility of utilizing for beneficial purposes some of the waste heat produced by a power plant. Some of the possible uses are for agricultural purposes, as a heat source for industrial processes, and to supply the space heating needs of large commercial and residential buildings. Another scheme for utilizing a greater portion of the chemical energy of the fuel in a power plant is cogeneration, the use of a boiler-generator system to generate both electricity and steam for industrial or other use. A schematic diagram of a cogeneration system is shown in Figure 2.4.

The example of energy transformation from chemical energy to electrical energy leads to the concept of efficiency, which is the ratio of useful energy output to energy input. As a general rule of thumb, it takes approximately 10,000 Btu of input chemical energy to produce a kilowatt-hour (3412 Btu) of electrical output. Some large modern power plants use less than 9000 Btu/kWh, depending on the size and design of the plant. It is for this reason that two different energy equivalents are used for the kilowatt-hour of electrical energy. In terms of energy end use, 1 kWh is equivalent to

3412 Btu, but approximately 10,000 Btu of fuel chemical energy is required to produce 1 kWh of electrical energy. The distinction is important when one is considering the impact of using electrical energy on energy resources. The overall efficiency of producing electricity from chemical energy is approximately (100)(3412/10,000) = 34%.

The energy balance is another important concept that is governed by the first law. For any open system, such as an industrial process, a building, or an entire industry, the first law requires an equality between the energy content of all inputs and the energy content of all outputs. The input consists of the energy in the fuels put into the system and the energy content of raw materials which are used in the system. The output consists of the energy content of materials leaving the system boundary and energy, including the so-called losses, leaving the system. This equality of input energy and output energy is a very useful tool in analyzing the percentage of input energy being used productively.

The Second Law of Thermodynamics

The second law of thermodynamics recognizes the quality of energy. It states that heat can never be completely converted into work in a cyclic process. While there is an energy equivalence of heat and work as required by the first law, the second law indicates that energy in the form of work is preferable to energy in the form of heat. Work can be completely transformed into heat but the heat cannot be completely transformed to work in a cyclic process. Thus, work is energy of a higher quality than heat because it is energy that is totally available. The various forms of energy can be ranked with respect to their availability for producing work. Work and electrical energy are high quality forms of energy, while thermal energy is of lower quality because it can only be partially converted into work.

The availability of heat is a function of the temperature of the heat source. Work is derived from heat by using a working fluid that undergoes a cyclic process starting from an initial state and returning to the same state after completion of the cycle. The ratio of the work output of a thermal cycle to the heat input is defined as the thermal efficiency of the cycle and is a direct measure of the fraction of heat added that can be converted into work. The theoretical maximum thermal efficiency that can be achieved is called the Carnot efficiency and is given by the expression $(1 - T_c/T_h)$ where T_h is the maximum temperature of the working fluid entering the cycle and T_c is the temperature of the working fluid after work has been extracted. Both temperatures must be measured on an absolute temperature scale.

As an illustration of the Carnot efficiency, the maximum temperature in a fossil fuel power plant is typically 2000°C (2000 + 273 = 2273 K) while the lowest temperature is typically 20°C (20 + 273 = 293 K). This gives a Carnot thermal efficiency of 100 (1 - 293/2273) = 87%. This is the theoretical maximum efficiency. Practical efficiencies lie considerably below this figure. The best efficiency achievable for a large power plant is slightly over 40%. The Carnot efficiency formula indicates that thermal efficiency can be increased by increasing the temperature of the heat source.

Since different forms of energy represent different qualities, the energy form used to meet a need should be matched to that need in a way that does not reduce unnecessarily the quality of the energy resources available. As a simple example, natural gas is burned in a home furnace with a flame temperature that exceeds 2000°F. The end use requirement of this high temperature flame is to heat air to 70°F. To a certain extent, it represents a mismatch of energy source and end use requirement to use a 2000°F energy source to meet the end use temperature requirement of 70°F. We have accepted this mismatch of energy source temperature and energy end use temperature for practical reasons, namely, to increase the rate at which energy can be transferred. Using a lower temperature heat source such as waste heat recovery and heat recovery from the environment by means of a heat pump would minimize the degradation of a high temperature heat source. In order to utilize our energy resources effectively, the quality of energy forms must be taken into account.

Entropy

One of the practical consequences of the second law is that as energy is transferred in a process, total energy is conserved but a degradation of quality of energy occurs. The thermodynamic property that measures the quality of energy is called entropy. A degradation of quality of energy results in an increase in entropy. It is desirable from the standpoint of the effective utilization of energy resources to design processes which minimize the increase of entropy. This can be accomplished to a certain extent by matching energy sources with the end use function to be achieved. In industrial processes it is sometimes possible to use energy cascading for this purpose. A simple example will illustrate the concept involved.

Consider a manufacturing plant which burns natural gas to heat parts in an annealing oven to 1000°F. The exhaust gas from the oven is vented to the atmosphere at above 600°F. The plant also requires hot water in a washing operation. Assume that the hot water is generated in a separate gas fired boiler. This situation presents the possibility of heating the hot water by using the exhaust heat from the oven. The arrangement would eliminate the need to use gas for water heating and thus reduce the utility bill. The energy in the gas that is burned in the annealing oven is then used in two stages. The first stage transfers some of the energy from the hot gases to the parts being annealed, thus reducing the temperature of the gas. The second stage transfers additional energy from the cooler gas to the water. The net result of this cascading is a better match of the quality of the heat source as manifested by its temperature to the end use heat requirements. Clearly, a cascading of heat utilization in this manner requires special equipment. Industries such as the chemical industry routinely use to advantage much more sophisticated energy cascading arrangements than that described in this example. In energy intensive industries, energy cascading is an essential component of good engineering design.

FURTHER ENERGY PRINCIPLES

The basic principles governing the manipulation of energy have been discussed, but all the details of energy-using processes cannot be fully covered here. However, some of the calculations which occur later are discussed here in order to demonstrate to the manager what is involved. The design of energy systems should be carried out by knowledgeable persons who can draw on such engineering principles, theorems, equations, and handbook information as may be required for the job in hand.

Energy Content of Hot Water

The energy contained in any material by virtue of its elevated temperature is determined by the specific heat of the material and is given by the formula

$$Q = C \times \text{mass} \times \text{temperature change}$$

where Q is the heat content (Btu) and C is the specific heat (Btu/lb-°F). The specific heat of water is 1 Btu/lb-°F. If, for example, a washing operation requires that 400 lb of water, brought in at 40°F, be heated to 180°F, used, and discarded, then the energy contained in the discarded water is

$$\frac{1\ \text{Btu}}{\text{lb-°F}} \times 400\ \text{lb} \times (180 - 40)\text{°F} = 56{,}000\ \text{Btu}$$

If the water is used at the rate of 400 lb/hr then energy is discarded, in the hot water, at the rate of 56,000 Btu/hr.

As another example, air is frequently exhausted from a heated building in order to control the buildup of odors or fumes. The exhausted hot air carries energy with it. The energy can be calculated using the same formula, and a specific heat for air* of 0.24 Btu/lb-°F. Since fan capacities are normally rated in cubic feet per minute (cfm), it is necessary first to convert cubic feet of

---

*Both the specific heat and the density of air (and other gases and gas mixtures) vary with temperature and pressure. The values given here are approximate, but are adequate for most energy management calculations.

air to pounds of air. The density of air at atmospheric conditions* is 0.075 lb/ft^3. The heat loss through an exhaust fan, when the exhausted air has been heated through a temperature range of ΔT°F, is given by

$$= 0.24 \text{ Btu/lb-°F} \times (\text{air flow}) \text{ ft}^3/\text{min} \times 0.075 \text{ lb/ft}^3 \times 60 \text{ min/hr} \times (\Delta T)\text{°F}$$

$$= 1.07 \times (\text{air flow}) \times (\Delta T) \text{ Btu/hr}$$

This is another useful energy formula, frequently found in handbooks.

Compressed Air

Air compressed to pressures up to about 100 psi (pounds per square inch) is used widely in industry. Compression of the air requires that work be done on it and the compressed air, therefore, is given potential energy. Thus, leakage or unintended uses of the compressed air constitute a waste or loss of the energy put into it during compression.

The work done in compressing the air depends on its temperature and the pressure to which it is compressed. Generally speaking, the lower the temperature of the air entering the compressor, the less energy is required for compression. Locating the compressor intake in as cool a location as possible is then frequently quoted as a measure which can save energy.

The detailed engineering calculations that go along with these two concepts are quite complicated and have no place in this book, particularly since the information regarding this process from an energy management point of view is readily available in an alternative form. Publications such as the EPIC Handbook series, American Society of Heating, Refrigerating and Air-Conditioning Engineering (ASHRAE) *Handbook of Fundamentals*, and the various engineering handbooks contain a wealth of information in tabular and graphical form relating to the energy conservation aspects of all commodities such as compressed air. Table 7.10, for example, gives the savings to

---

*See footnote, p. 34.

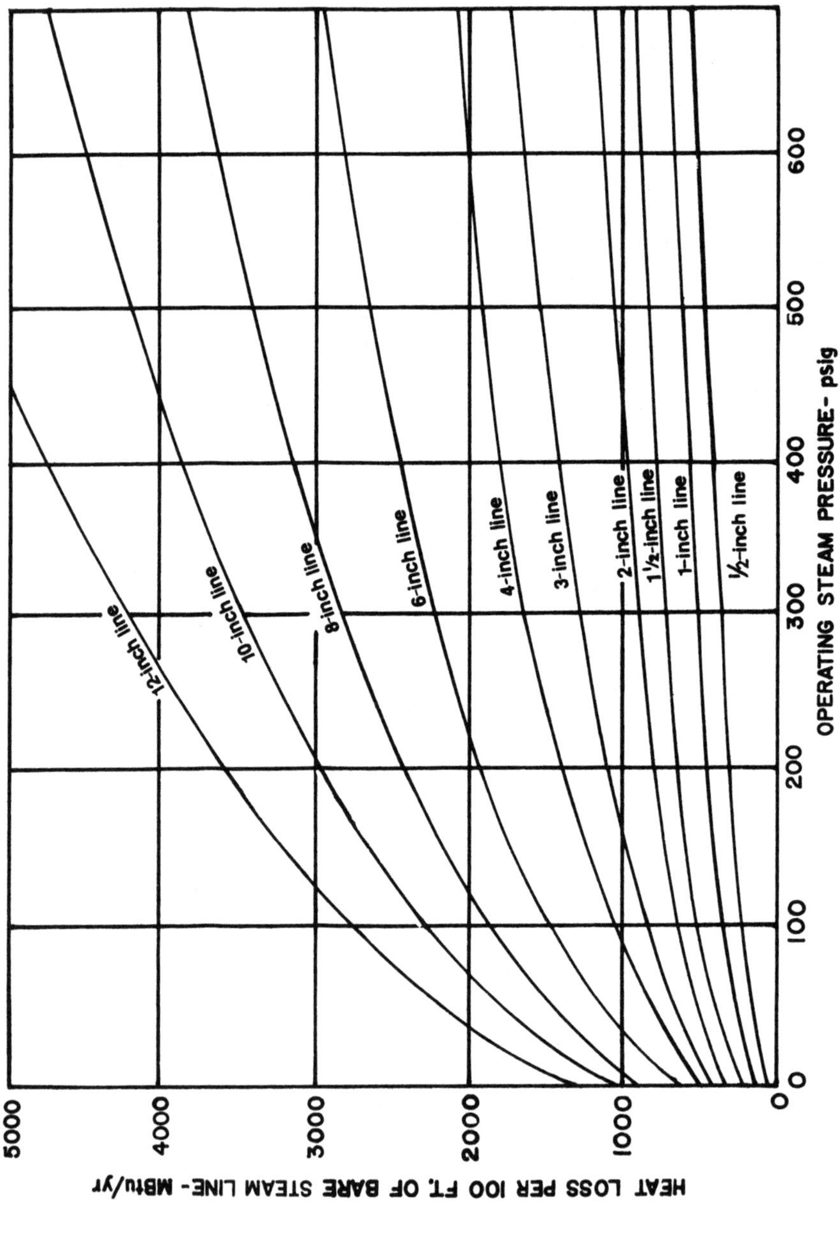

Figure 2.5 Heat loss from bare steam lines. (Reproduced from Ref. 2, pp. 3-19)

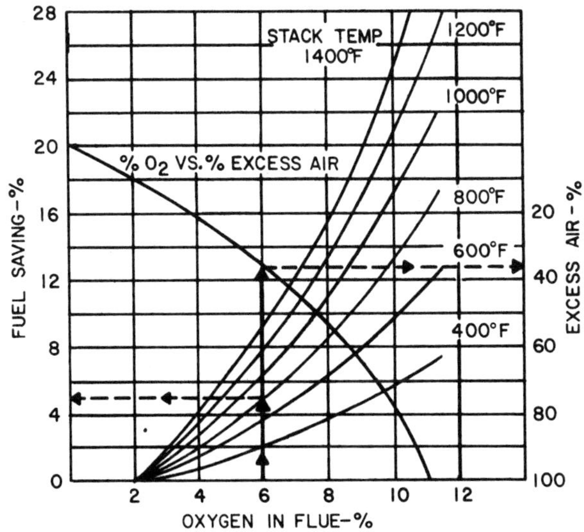

Figure 2.6. Boiler fuel savings available by reducing excess air. (Hydrocarbon gaseous fuel saving available by reducing excess air to 10%.) Reprinted from Ref. 2, pp. 3-47, with permission from Bailey Controls Company.)

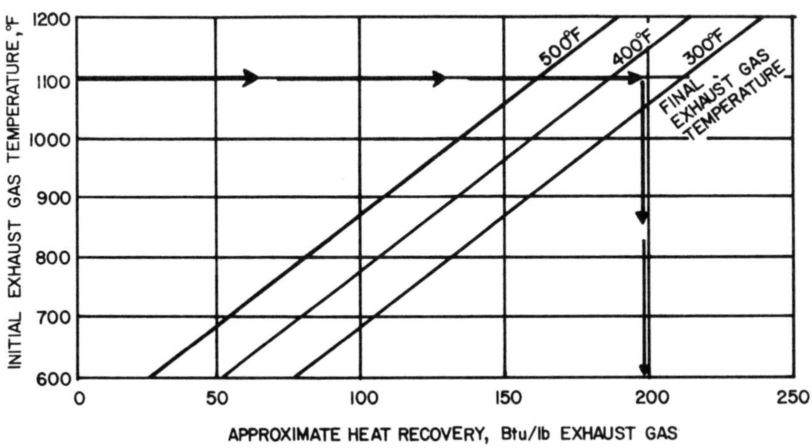

Figure 2.7. Heat recovery from gas and diesel engine exhaust gas (based on 2% heat loss to atmosphere). (Reprinted from Ref. 2, p. 3-41, with permission from Deltak Corporation.)

be obtained by reducing the temperature of the intake air to a compressor and is reproduced from EPIC. Other energy conservation matters covered in the same handbook include various aspects of the installation, use, and maintenance of steam systems (Fig. 2.5), the adjustment and firing of boilers for greater efficiency (Fig. 2.6), and recovery of waste heat (Fig. 2.7). As the reader looks further into the many engineering opportunities to improve on the use of energy, a file of useful handbooks and handbook material will be accumulated. By using this resource many of the commonly occurring engineering questions in energy management can be better understood at least, even if not fully resolved.

Recovery and Use of Normally Wasted Heat

Waste heat recovery is another area in which a great deal of engineering knowledge is required to carry out design and application. Handbook information is again useful in this area, particularly regarding the effective use of the waste heat; for example, EPIC contains information on the beneficial effects of preheating boiler combustion air (Fig. 2.8). Here the consideration is for the

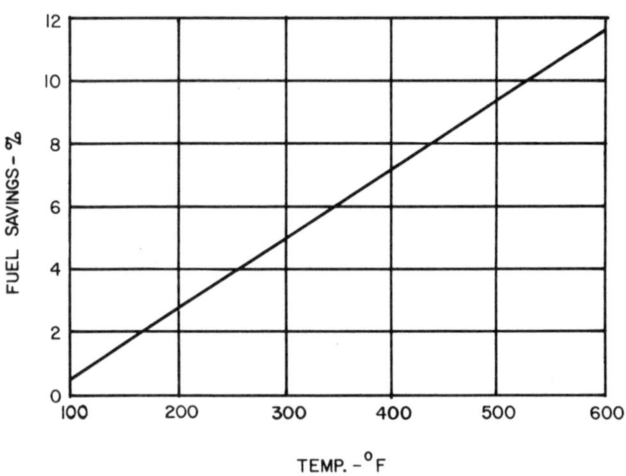

Figure 2.8. Boiler fuel savings from preheating combustion air. (Reproduced from Ref. 2, p. 42.)

combustion air in a boiler. The air must be raised to the temperature resulting from the burning of the fuel. If cold (ambient) air is used more energy is required than if heated air is used (because of the specific heat of air already discussed). The order of efficiency improvement to be expected from various preheat temperatures is given in the graph for easy use. The preheating is often achieved through recovery of heat from an exhaust stream (air, water, or other fluid).

A further source of information in this case is the performance specifications literature available from manufacturers and vendors of waste heat recovery equipment (Fig. 2.9). Again, a file of useful information can be easily and quickly assembled by anybody active in this area.

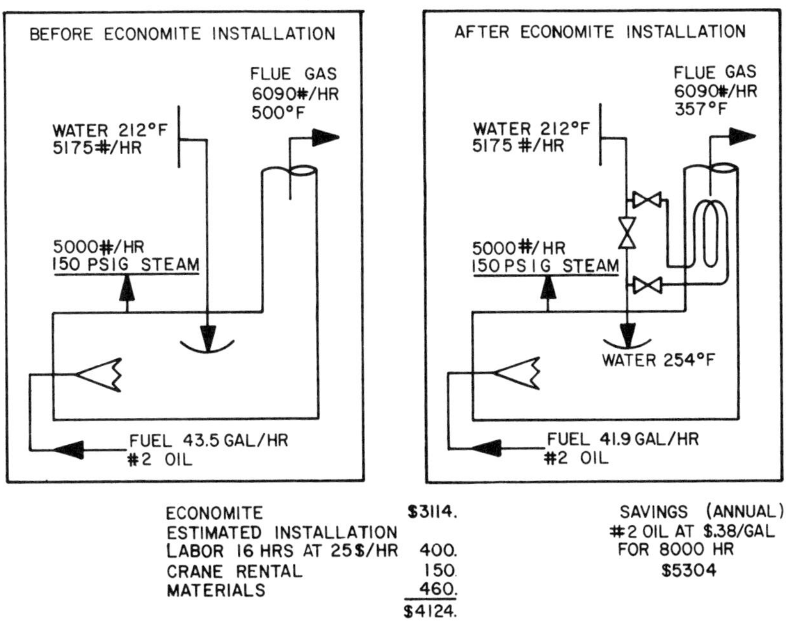

Figure 2.9. Flue gas heat recovery equipment. Typical savings from Economite installation (Economite 150 18 10 404). (Reprinted with permission from Economite Fuel Economizers for Small Boilers, E-Tech, Inc., Tulsa, Okla., 1976, p. 2.)

Heat Conduction

Heat conduction is a recurring and important subject in energy management. The ability to conduct heat is a necessary property of the tube walls in boilers and heat exchangers; it is an undesirable property of thermal insulation. In the latter case the insulating material is designed to have low thermal conductivity--the ability of a material to conduct heat.

The equation for heat flow by conduction from a body at a temperature T_1°F to a body at temperature T_2°F is

$$Q = UA(T_1 - T_2) \qquad \text{Btu/hr}$$

where U is the thermal conductance measured in Btu/hr-ft^2-°F and A is the cross-sectional area (ft^2) of the path of heat conduction. Since U is inversely proportional to the length of the heat conduction path, the walls of the tubes in a heat exchanger are made as thin as permitted by other considerations and the depth of insulation in an attic is made as great as economically feasible. In the case of building insulation the conducting property of the insulation is frequently given as an R-value (the reciprocal of the U-value) which emphasizes its heat resisting qualities. Then

$$Q = \frac{A}{R}(T_1 - T_2) \qquad \text{Btu/hr}$$

Documentation

In view of the wide range of information sources available, the energy manager would be well advised to insist that all technical input to an energy management decision be fully and accurately documented. As new information and improved equipment become available, changes in the technical evaluation of a project can be readily made.

FURTHER ENERGY RELATIONSHIPS

Power is a concept normally associated with energy in the form of work and is the rate at which work is done. Energy use is related to power by the expression: energy = average power × time. This relation between energy and power is important in conducting an energy audit and application of this concept is discussed in Chapter 6.

Power is also important in determining the total cost of energy because electrical utilities frequently charge for the customer's power requirement as well as for the total energy used. The charge associated with the power requirement of the customer is called the demand charge.

The Rating of Equipment

It is the power rating which is usually indicated on energy using equipment. A gas water heater, for example, may be rated at 550,000 Btu/hr. This is the rate of energy use for which the equipment is designed. If it runs at rated conditions and runs for one hour, the energy used during that time will be 550,000 Btu.

Electric power ratings are given in watts or, more frequently, in kilowatts (kW). A power level of 1 watt is equivalent to energy transfer at the rate of 1 J/sec, but since the joule is a unit that is rarely used in engineering practice, electrical energy is usually measured in watt-hours. One watt-hour is the energy transferred in one hour when the power level is 1 watt. One kilowatt-hour (kWh) is the energy transferred in one hour when the power level is 1 kW. Electric motors are usually rated in terms of horsepower (hp). This is a measure of the maximum rate at which the motor can do work. Reference to Table 2.1 will show that 1 hp is equivalent to 746 watts or 0.746 kW. A 5-hp motor operating at full load for 1 hour will, therefore, deliver 3730 watt-hours (or 3.73 kWh) of energy. Since the nameplate horsepower is the rating of the motor output, the power delivered to the motor from the electrical system will be more than 5 hp to account for losses in the motor. The output power must be divided by the efficiency to obtain the input power. For small motors, up to about 5 hp, full load efficiencies will vary from about 60% to 80%. For larger motors, full load efficiencies may be greater than 90%. These figures should be considered as typical values since the actual efficiency will depend on the details and objectives of the motor design.

Table 2.3. *Electrical Power Formulas*

Single-phase circuits	Power = voltage × current × power factor
Three-phase circuits	Power = 1.732 × voltage × current × power factor

Other electric equipment may be found to have a rating specified by voltage and current instead of horsepower. For alternating-current equipment the power rating can be found by using one of the formulas shown in Table 2.3. The power factor is a measure of how effectively the electric current is being used to produce an effect. It depends on the components of the particular electric currents involved. For heating elements and incandescent lights power factor can be taken as 1.0. For so-called inductive circuits, such as welders, power factor will be somewhere between 0 and 1.0. For many induction motors the power factor will range from 0.4 to 0.8, depending on the load on the motor.

Thermostatic and Intermittent Controllers

For equipment which is controlled by a thermostat or a timer care must be taken to use the time the equipment is actually running at rated power in calculating the energy used. This will be less than the time the equipment is in actual use since the thermostat or controller will automatically shut off the circuit or fuel supply in accordance with some predetermined program or criterion.

Variable Loads

From a practical point of view, electric motors present a difficult problem when one is trying to determine energy use from the power rating on the nameplate. The rating of an electric motor is a measure of output capability. However, if the load being driven requires less than the rated horsepower the motor will deliver only the horsepower required and not the rated horsepower. In many applications the load on an electric motor may vary continuously because the load being driven by the motor is varying. The drive motor on an escalator is an example of a variable load. With no one on the escalator, the motor load is very light and is essentially equal to the power required to overcome the friction of the moving escalator system. When the escalator is fully loaded with passengers, the motor load is very heavy and may be close to the nameplate rating of the motor if the motor has been sized properly.

In such cases of variable load, in order to calculate energy used an estimate must be made of the *average* power level at which the motor is working. Making such an estimate requires a certain degree of skill and experience but the need for it must be fully understood in order to make a correct determination of energy consumption.

REFERENCES

1. J. B. Marion, *Energy in Perspective*, Academic, New York, 1974.
2. U.S. Department of Commerce, *Energy Conservation Program Guide for Industry and Commerce*, NBS Handbook 115, U.S. Government Printing Office, Washington, D.C., 1974.

SUGGESTED READING

J.S. Doolittle, *Energy: A Crisis, A Dilemma, or Just Another Problem,* Matrix, Champaign, Ill., 1977.

R.C. Dorf, *Energy, Resources and Public Policy,* Addison Wesley, Reading, Mass., 1978.

J. C. Fisher, *Energy Crises in Perspective,* Wiley, New York, 1974.

J. C. Sawhill, ed., *Energy Conservation and Public Policy,* Prentice-Hall, Englewood Cliffs, N.J., 1979.

U.S. Department of Commerce, *Energy Conservation Program Guide for Industry and Commerce,* NBS Handbook 115, Supplement 1, U.S. Government Printing Office, Washington, D.C., 1974.

U.S. Department of Commerce, *Waste Heate Management Guidebook,* NBS Handbook 121, U.S. Government Printing Office, Washington, D.C., 1976.

Chapter 3

BASIC ECONOMIC CONCEPTS OF SUPPLY AND DEMAND

SCARCITY AND RESOURCE UTILIZATION

A basic fact of economic life is that of scarcity. Indeed, the existence of business firms can be traced to the condition of scarcity. If resources and the products created from them were not scarce, each individual would be in a position to supply his or her own needs. In actuality, however, various types of business organizations are established which combine scarce resources and produce goods and services which are made available to the final user.

The study of how individuals and groups decide to use scarce resources in the satisfaction of human wants is the discipline of economics. Therefore, a review of some basic economic concepts will help us understand the energy problem and provide a framework for considering the responses of a firm or individual to changing energy prices and scarce supplies. Also, it will reveal alternative approaches to global energy "shortage" issues.

Resources or "factors of production" are classified into four broad categories in the study of economics:

1. *Land* all natural resources such as agricultural land and underground gas and petroleum reserves
2. *Labor* human effort "constructively" applied
3. *Capital* man-made resources used over and over in the production process (i.e., plant and equipment)

4. *Management* a special type of "labor" performed by managers in a firm to combine the other input resources so as to create useful goods and services

Owners of these factors of production receive returns for the use of their resources which are called economic rent, wages, interest, and profit (for owner-managers), respectively. In the next chapter, this traditional list of resources will be refined to include energy and specific goods used in production systems or other operating systems.

Resource scarcity is easily verified by casual observation. Human wants for goods and services seem to be insatiable, while the quantities available are always limited. Scarcity thus poses a problem: If resources are not abundant enough to allow us to do everything, how should those we have be used? Which of the unlimited wants should be satisfied? In principle, the limited resources should be used in a way to produce the greatest amount of satisfaction. This basic truth should apply to the individual consumer who is pondering the acquisition of a new television set, to the school board which is deciding on the location of a new school building, and to the industrial corporation manager who is allocating capital among alternative investments.

The nature of the economic problem is illustrated in Figure 3.1. Scarce resources cannot be stretched to cover all human wants. If one group of wants is met through the use of the resources which are limited at a moment of time,* other wants will go unmet. If, however, we can agree that satisfaction should be maximized (and if we agree that such an intangible as "satisfaction" is meaningful and measurable), we have a basis for ordering and directing economic activity.

There are two basic approaches to determine what is satisfying. In an authoritarian society, some central authority may direct the use of resources and the allocation of goods and services.

---

*Note that over time the quantity of resources can and usually does change. Also, changing technology may mean that a given stock of resources can perhaps be used to satisfy more wants over time.

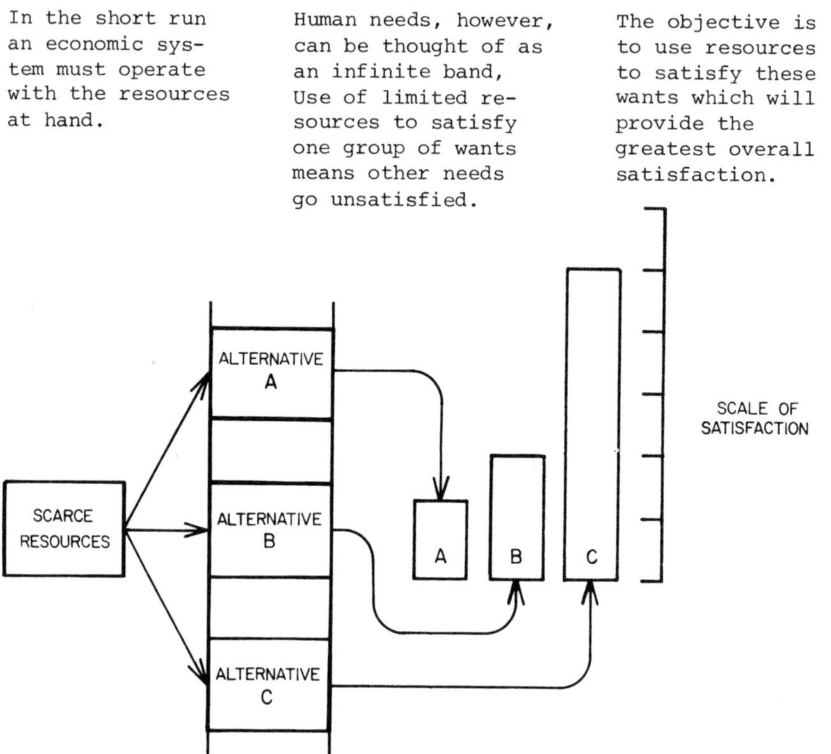

Figure 3.1. The resource allocation process: in the situation illustrated here, the best use of society's resources is alternative C which results in the highest level of satisfaction. Source: Adapted from Moore, John R. et al., *Economics: Principles, Problems, and Perspectives* (Boston: Allyn and Bacon, 1966, p. 9).

In a free enterprise society, decisions are left to individuals and manifest themselves in the prices which are paid for products and resources. In fact, almost all economic systems involve a blending of these two allocative mechanisms. In the U.S. economy, the emphasis is on the price system and individual decisions, but there are elements of "collective" decision-making when governmental units decide how much to spend on highways, weapons, education, and courthouse construction. On the other hand, the role of central direction is clearly greater in nations such as the Soviet Union, but the price system also has a role to play.

MARKETS AND PRICES

In economic systems like that of the United States where resources are allocated on the basis of price, the price established in any market at any time depends on two factors:

1. The intensity of desire for a product or factor of production and the ability to pay for it
2. The willingness and ability of the producer or seller to provide the item

The first variable is referred to as "demand" and the second as "supply."

The concept of demand embraces both desire and the ability to pay so it is really "effective demand." The number of units of a product or service an individual will buy is influenced by the amount of income the buyer receives, the price of the item, the price of close substitutes, the taste and preferences of the consumer, and perhaps the amount of promotional activity and other factors.

Demand in the short run is defined as a schedule of the amounts an individual would be willing to purchase at all possible prices at a moment in time. If, for example, we consider the demand of Mr. Able for gasoline this week (a moment in time), and assume that his income and other factors remain constant, his demand might be like that shown in Table 3.1 and Figure 3.2. It is quite usual to find this general type of behavior, where the number of units wanted goes down as price goes up. In fact, this situation is so common that it has led to a basic law of economics which holds that there tends to

Table 3.1. Hypothetical Gasoline Demand

Price	Number of gallons purchased (week 1)
$2.00	0
1.50	5
1.00	10
0.50	15

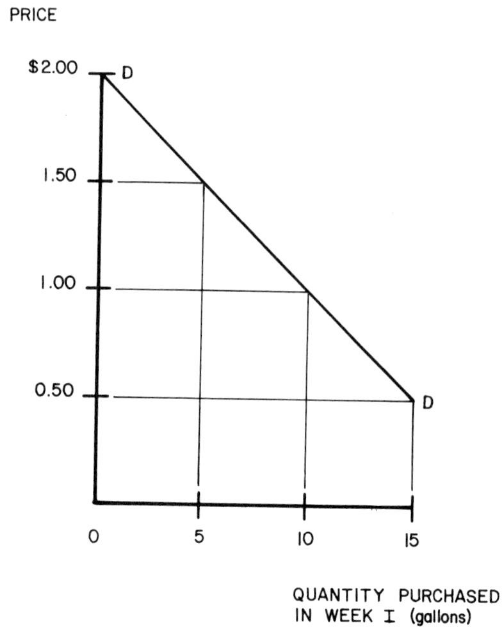

Figure 3.2. Hypothetical gasoline demand.

be an inverse relationship between the price of an item and the quantity of that item demanded. This relationship can be expressed with symbols as follows:

$$Q_d^g = f(P^g; Y, P^s, T, A)$$

This equation indicates that the quantity of gasoline demanded (Q_d^g) is a function of the price of gasoline (P^g), assuming those factors appearing after the semicolon--income (Y), the price of substitutes (P^s), taste (T), and institutional factors such as advertising (A)--are constant. In this illustration, and in general, the demand curve slopes down from left to right. However, the price-quantity relationship is not necessarily "smooth" or represented by a straight line.

A logical analysis will indicate why this inverse relationship exists between price and quantity demanded. First, the time period under consideration is defined as the short run, so that factors such as income and the prices of other products are assumed to remain constant. This assumption is critical to understanding the law of demand, because, if other product prices or income changes, the consumer's behavior may reflect a substitution effect or income effect.

To illustrate the substitution effect, let us assume that the price of electricity is constant and the price of natural gas increases. If other things are equal, a substitution of electricity for natural gas and a reduction in natural gas sales will occur. Conversely, if the natural gas price decreases while the price of electricity remains constant, natural gas sales will increase as natural gas is substituted for electricity. The degree of similarity between products which are substitutes will affect the actual amount of substitution. If a product has no close substitutes, then price movements either upward or downward may have little effect on the quantity sold.

However, in situations where the substitution effect is not operative, demand may be modified by an income effect. If we assume that the consumer's income is limited in the short run, then the consumer's buying power is also limited. The amount which can be spent will depend on the fixed sum of dollars available from income, if we ignore the possibilities of using savings or incurring debt. However, the dollar amount spent will be a function of both price and quantity. If price goes up and income remains constant, the quantity purchased must go down. Thus, the income effect tends to cause the quantity demanded to go down as prices go up. In reality, both substitution and income effects are operative at the same time, and the consumer must make decisions relating to several products or services instead of one.

The price-quantity relationships discussed above pertained to individuals. Similar relationships can be described for groups of individuals who have "wants" for a particular product. The sum of the quantities which individual consumers would purchase at all

Table 3.2. Hypothetical Market Demand Schedule for Gasoline

	Number of gallons demanded by				
Price	Mr. Able	Ms. Baker	Mr. Carr	Mrs. Delta	Total
$2.00	0	2	7	1	10
1.50	5	4	10	12	31
1.00	10	7	16	17	50
0.50	15	7	16	18	56

possible prices is the market demand. A hypothetical short-run market demand schedule for gasoline is shown in Table 3.2. Using the data in the price and total demand columns, a graphical demand curve could be constructed. Recall that this demand curve assumes many factors remain constant. These additional influences on demand (e.g., income) will become more important as the time period considered lengthens.

EMPIRICAL MEASUREMENT OF DEMAND

The preceding discussion of demand relationships has been somewhat abstract, but decision makers need to obtain estimates of price-quantity relationships in the real world. Knowledge of such relationships is valuable to the business manager who is developing product pricing and it is vital for government officials who establish public policies. For example, suppose the Administration and the Congress are interested in promoting the conservation of gasoline. The methods for achieving this result and the motivation of consumers will be quite different if the price-quantity relationship is that shown between prices of $0.50 and $1.00 in Table 3.2 or if it is similar to the one shown between prices of $1.50 and $2.00! How can this relationship be established and what other nonprice factors may be important?

There are essentially three approaches: (1) the intuitive, (2) the experimental, and (3) the statistical or "econometric" approach. The intuitive approach involves a logical analysis of changes in demand due to substitution and income effects. If we know that few substitutes exist for a product and that a small fraction of the consumer's income is spent on that product, it is likely that price changes per se will have little effect on the number of units demanded. On the other hand, if there are close substitutes for a product and/or the consumer spends much of his or her income for the item, significant changes are likely in the quantity demanded as price changes.

The experimental approach seeks to develop a number of price-quantity relationships via controlled experiments. Assume that Exxon is interested in gaging potential consumer response to a new grade of gasoline. They might seek out two market areas which are similar in size, population, and income levels and try to market the new product at a price of $0.65 in one area and $0.80 in the other market. If the sample market areas are properly selected, it may be possible to generalize about global price-quantity relationships from these limited observations. This approach is most commonly used when dealing with new products.

The third method of empirical demand estimation, the statistical or econometric approach, is used for products which have some market exposure. It requires the assembly and statistical analysis of data on past price-quantity relationships and the relationship of demand to other factors.* A typical result of such an analysis is an equation which expresses changes in quantity, the dependent variable, as a function of price and income. An example of such an econometric demand equation is

$$Q_d^x = 1700 - 54P_x + 1.02Y$$

where Q_d^x is the quantity of product x demanded at the price P_x. If the price of product x was $1.00 and income was $100, the estimate of quantity demanded based on this equation would be 1748 units.

---

*An elementary discussion of various approaches to demand forecasting may be found in Chapter 5 of Brigham and Pappas (see "Suggested Reading").

ELASTICITY OF DEMAND AND SUPPLY

Basic Concept of Elasticity

The concept of elasticity is used to indicate certain relationships between two economic variables. In general, it is a measure of the change in a dependent variable in relation to a change in an independent variable. It is commonly used to determine the change in demand (or supply) for a given change in price, and this index is called the price elasticity of demand (or supply). When considering price-quantity relationships, price elasticity (E_p) is the ratio of a change in quantity (relative to the average quantity) to the change in price (relative to the average price). The equation for price elasticity is as follows:

$$E_p = \frac{\Delta Q/\bar{Q}}{\Delta P/\bar{P}} = \frac{\Delta Q}{(Q_1 + Q_2)/2} \div \frac{\Delta P}{(P_1 + P_2)/2}$$

It measures the "arc" price elasticity of demand, or that based on changes between two price points on a demand curve.*

The concept and significance of elasticity can be best illustrated with a simple example. Suppose the price-quantity relationships are as shown in the first two columns of Table 3.3 and in Figure 3.3. The calculated price elasticities of demand are shown in the third column of Table 3.3. These coefficients of elasticity indicate the fractional change in quantity for a given fractional change in price. Price elasticities of demand are classified in five categories as follows:

1. Perfectly inelastic demand (coefficient of elasticity is zero)
2. Relatively inelastic demand (coefficient of elasticity is between zero and one)
3. Unitary elastic demand (coefficient of elasticity is one)
4. Relatively elastic demand (coefficient of elasticity is between one and infinity)
5. Perfectly elastic demand (coefficient of elasticity is infinity)

---

*Point price elasticity (E_p), or the responsiveness of quantity to infinitesimally small changes in price, is found by using the equation, $E_p = dQ/dP \times P/Q$.

Table 3.3. Hypothetical Demand Schedule

Price	Quantity	Elasticity*	Total revenue	Marginal revenue
$10	1		10	-
		6.33		
9	2		18	8
		3.40		
8	3		24	6
		2.14		
7	4		28	4
		1.44		
6	5		30	2
		1.00		
5	6		30	0
		0.69		
4	7		28	-2
		0.47		
3	8		24	-4
		0.29		
2	9		18	-6

*Since price and quantity are usually inversely related, the number representing price elasticity carries a negative sign. For our purposes we can safely ignore the sign.

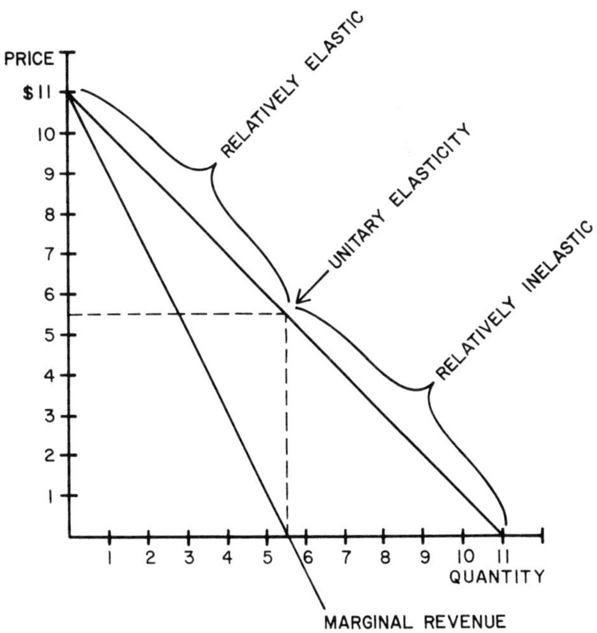

Figure 3.3. Hypothetical demand curve.

In our example, when the price went from $10 to $9 the fractional increasing quantity demanded was 6.33 times the fractional decrease in price, so this is a situation of relatively elastic demand. On the other hand, if the price rose from $3 to $4, the fractional change in quantity would be smaller than the fractional change in price, the ratio would be 0.47, and demand would be relatively inelastic.

Elasticity is not equal to the slope of the demand curve; for example, the slope of the demand line in Figure 3.3 is constant while elasticity changes as one moves along the curve. This is true because the coefficient of elasticity is the product of the slope of the demand curve *and* the ratio of price to quantity. At each price, the price/quantity ratio is different.

Elasticity does tell us how responsive the quantity demand is to price manipulation or to other induced price changes. Also, it tells us something about how other important economic variables such as total revenue and marginal revenue* will be influenced by price change. For example, when demand is relatively elastic, a decrease in price results in a greater total revenue (price × quantity). When demand is relatively inelastic, a decrease in price reduces total revenue. Of course, these relationships are reversed when prices are increased. If unitary elasticity exists, changes in price will not affect total revenue. These relationships are shown numerically in Table 3.3.

Marginal revenue is always positive if demand is relatively elastic, but if demand is releatively inelastic, marginal revenue is negative. When elasticity is unitary, marginal revenue is zero. These relationships are shown in Table 3.3 and Figure 3.3.

Long-term Demand Shifts and Elasticity

Over a longer period of time, it is unrealistic to assume that "other things will remain constant." As tastes, incomes, and other prices change, the simple relationships outlined above may not hold

---

*Marginal revenue is the amount of additional revenue which results from producing and selling one more unit of a product.

true. In the preceding illustration, Mr. Able wished to purchase 15 gallons of gas at a price of $0.50 per gallon *if* everything else was held constant. However, if he receives a promotion and a substantial salary increase, he may purchase a new, larger automobile and then buy 18 gallons of gas per week at the $0.50 price. The result is a shift in the entire demand curve to the right. Such a shift in the demand curve (to the left or right) which results from a change in one or more underlying demand variables (demand determinants) is referred to as a "change of demand" as distinct from a "change in the quantity demanded." The latter term was used to describe changes along a given demand curve as the result of price changes. A change in demand is illustrated in Figure 3.4.

The general concept of elasticity can also be applied to longer-term demand relationships. In the equation for price elasticity, the independent variable was price. In dealing with an expanded range of demand determinants, other independent variables such as income, product promotion expenditures, and price changes in other products may be used to determine the shift in demand for a given product. These elasticities are defined as follows:

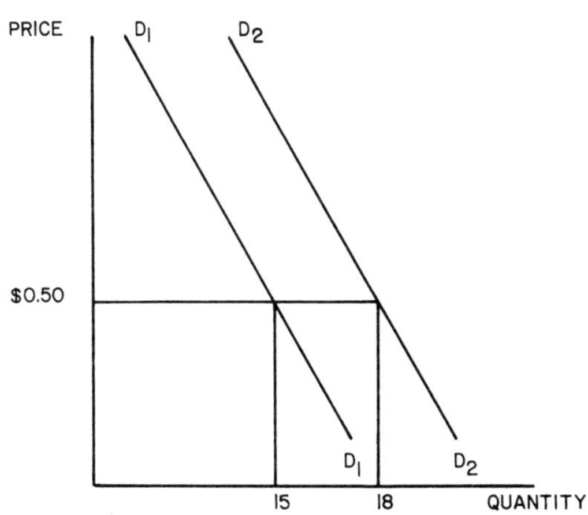

Figure 3.4. Hypothetical gasoline demand curve shift.

$$\text{Income elasticity} = \frac{\text{fractional change in demand}}{\text{fractional change in income}}$$

$$\text{Promotional elasticity} = \frac{\text{fractional change in total revenue}}{\text{fractional change in advertising expenditure}}$$

$$\text{Cross-elasticity} = \frac{\text{fractional change in demand for product Y}}{\text{fractional change in price of product X}}$$

Among these relationships, income elasticity and cross-elasticity are most important in energy management. Historically, as incomes have increased, the demand for energy-using devices has increased and total energy demand, especially for electricity, has increased steadily. Cross-elasticity is becoming more important in relation to energy because the relative prices of various forms of energy are shifting rapidly and the substitution of one form for another is often influenced significantly by price changes.

Applications of Elasticity Concepts to Energy Developments

The concept of price elasticity of demand can be used to analyze changes related to energy. To illustrate, assume that (1) the elasticity of demand for petroleum products in the United States is 0.1, (2) all petroleum products are imported, (3) the consumption was 15 billion barrels per year in 1973 at a price of $2.50/bbl, and (4) a cartel of producers forces the price to $10.00/bbl in 1974. In this situation, the low coefficient of elasticity indicates that demand is very inelastic and will not change much, even though the price increases by a large amount. It is logical to expect gasoline price increases to have a small effect on demand, since there are no close substitutes and the fraction of each consumer's income spent on gasoline is relatively small. In terms of total spending for gasoline by all consumers in the nation, the level increased from $37.5 billion in 1973 to approximately $133 billion in 1974.* Thus, if the

---

*The 1974 value is found by solving for Q_2 in the equation:

$$E_p = \frac{Q_1 - Q_2}{(Q_1 + Q_2)/2} \div \frac{P_1 - P_2}{(P_1 + P_2)/2}$$

demand for energy is relatively price inelastic, significant increases in total spending for energy will occur as prices increase, with a corresponding decrease in spending for other goods or services.

In some operations which utilize energy, one type of fuel may be substituted for another, and this will occur as relative prices change. For example, natural gas and coal are close substitutes as fuel for boilers. Assume the price of coal increases by 20% due to stricter federal regulations on environmental protection and mine safety, while natural gas prices remain constant due to continuation of federal price ceilings. Since fuel costs are a significant fraction of total costs in thermal energy generating facilities and fuel substitution is relatively easy, cross-elasticity will probably be high and the demand for natural gas will increase significantly, while the demand for coal will fall sharply. The implications of such shifts are great for many economic "actors." For example, the investment required to switch from coal to gas may be significant; also, the total revenue of coal producers may decline despite the coal price increase.

Supply Considerations

To this point, only the demand side of the pricing process has been discussed. However, most of the types of statements made about demand could and should be made about supply.

Supply is defined as a schedule of the amounts that would be offered for sale at all possible prices at a particular time. Thus, in the short run, the focus is again on a price-quantity relationship. Over a longer period of time, factors such as the prices of the labor, capital, and raw material inputs and the available technology will affect supply. Changes in any of these long-term factors lead to a change in supply--that is, a shifting upward or downward of the supply schedule--rather than a change in the quantity supplied (a movement along a given supply schedule as a result of a change in price).

In general, there is a direct relationship between price and quantity, and this relationship can also be expressed in terms of supply elasticity.* A number of factors affect supply elasticity, but time is the most important. In the immediate time period, supply may be completely inelastic. That is, it may be impossible to increase the supply of a product or resource at a given moment no matter what price inducement is offered. The longer the response period, however, the more likely it is that supply can be increased. Of course, the long-run supply elasticity will depend on the nature of the product and the inputs which go into it.

A typical supply curve is shown in Figure 3.5. The line S_1S_1 is drawn to show a short-run supply situation, while S_2S_2 shows a shift in the supply curve which could occur over time. Such a shift might be due to depletion of a natural resource which required higher-cost secondary or tertiary recovery techniques, an increase in the cost of labor inputs, or tax increases. As in the case of

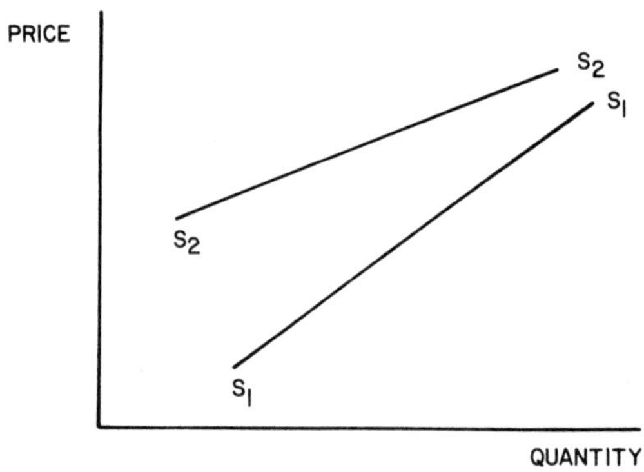

Figure 3.5. Hypothetical gasoline supply curve shift. (Note: As in the case of demand, the market supply can be built up from the willingness of individual producers to supply more at a range of prices.

---

*Supply elasticity is defined as the fractional change in quantity supplied for a given fractional change in price. Supply responses may be classified as perfectly inelastic, relatively inelastic, unitary, relatively elastic, or perfectly elastic, depending on the coefficient in the same manner as the price elasticities of demand.

demand, all factors but price are assumed to be constant in the short run, while everything may vary in the long run.

SUPPLY AND DEMAND EQUILIBRIUM

Price Determination

By combining the basic concepts of demand and those of supply, we can explain how prices are set and how markets perform. Hypothetical demand and supply data are shown in Table 3.4 for a given market in the short run. The quantities which would be demanded and the quantities which would be supplied at a range of prices are included, but not which price will prevail. However, the prevailing price will tend to be the equilibrium price, where the quantity demanded is equal to the quantity supplied. In this example, the equilibrium price would be $1.00, as shown in Figure 3.6.

To understand why this must be the market price, consider what would happen if some other price were the initial price. Suppose that the price is $1.50; at that price, buyers would want only 5 units, but sellers would like to provide 15 units. There would be a surplus, since some sellers could not find buyers. These sellers would be willing to accept a lower price and thus prices would tend to fall toward the $1.00 equilibrium level. On the other hand, if the price were initially at $0.50, the amount buyers would like to buy would be 15 units and the amount sellers want to sell would be only 6 units. There would be a scarcity, with consumers willing to pay more to satisfy their desires. The result would be competition

Table 3.4. Hypothetical Short-run Demand and Supply for Gasoline

Price	Quantity demanded	Quantity supplied
$2.00	0	20
1.50	5	15
1.00	10	10
0.50	15	6

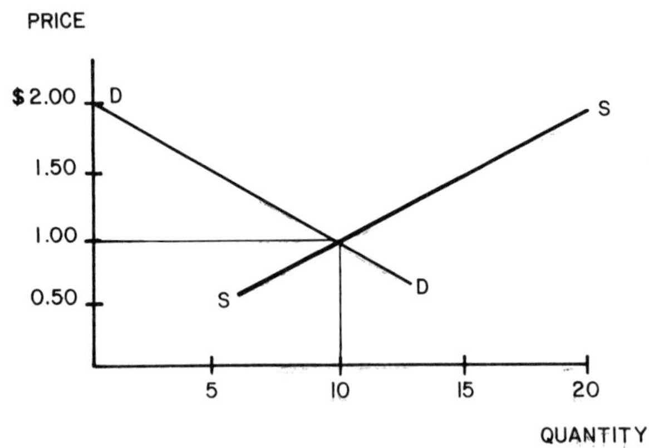

Figure 3.6. Hypothetical gasoline demand and supply curve (short run).

for the available goods which would tend to force the price upward toward market equilibrium at the price of $1.00.

Caution is required when considering price setting in the market through the interaction of supply and demand. The process described is one which tends toward a state of balance or equilibrium. However, equilibrium at a particular price may or may not be desirable, or it may be desirable for some and undesirable for others. What is *desirable* involves a value judgment. What we have described is a process which leads to a result. The distinction between "what is" and "what should be," or between descriptive and normative, should be clearly made.

For example, people often say: "You can't interfere with supply and demand." In context, this usually conveys the meaning that there is something right about a particular price or that the equilibrium process leads to a good price. Thus, the implication is that the natural forces which created that price should not be changed in any fundamental, long-term fashion. However, nothing could be further from the truth for one of the parties whose individual demand has been toted up to establish the market demand and may be one who is seeking to influence the price. That party may be a William Cunningham who bids up the price of tobacco due to his "greed," or it may be the government seeking the good purpose of income stabilization for one segment of the farming community.* Any shift in either the demand curve or the supply curve will cause the price to change in most cases. Thus, it is possible to interfere with supply and demand. More specifically, one can alter market prices through manipulation of the underlying conditions affecting supply and demand. Indeed, much of our discussion of energy-related issues will focus on just this point.

In summary, prices change when demand and/or supply curves shift. Changes in income, taste, advertising, and the price of other products may cause a shift in the demand curve as parties move into or out of the market and thereby change the quantities demanded at various prices. Supply curves shift because of changes in underlying factor prices and/or technology which cause suppliers to enter or leave the market or change their willingness to supply at given prices. The effects of such shifts in demand and/or supply will depend on elasticities. For example, if demand is perfectly elastic, as shown in Figure 3.7(a), a shift in the supply curve will increase the quantity sold but will have no effect on price; this is true when the shift involves either an increase in supply (leftward shift) or a decrease in supply (rightward shift). On the other hand, given a perfectly inelastic supply curve, the impact of demand shifts will be felt in terms of price alone and quantities will not change at all; this is illustrated in Figure 3.7(b).

---

*This example is not intended to imply that the actions of private traders are likely to be bad and that governments always wear "white hats." Indeed, one could readily think of examples in which the roles probably would be reversed. Most Americans would regard the activities of the OPEC member governments, for example, as being motivated by greed. Fortunately, or unfortunately, the market makes no moral judgments.

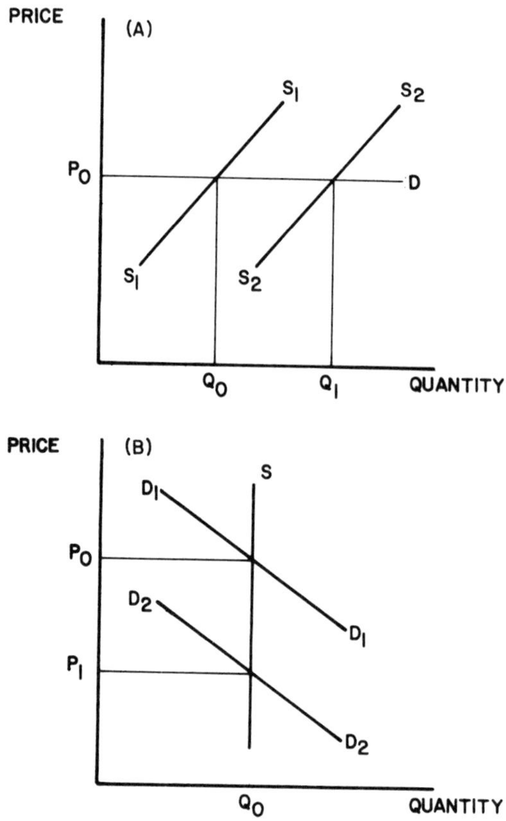

Figure 3.7. Demand and supply curve shift effects. (a) When demand is perfectly elastic, supply shifts result in no change in equilibrium price. (b) When supply is perfectly inelastic, demand shifts result in no change in equilibrium quantity.

Partial versus General Equilibrium

The preceding discussion of the role of demand and supply in establishing the equilibrium price and quantity in a given market tells only part of the story, so it is referred to as partial equilibrium analysis. In the marketplace, changes which result in demand or supply shifts in one market often cause shifts in other markets. For example, the demand curve for electrical energy might shift to the left as a result of the installation of a heat pump.

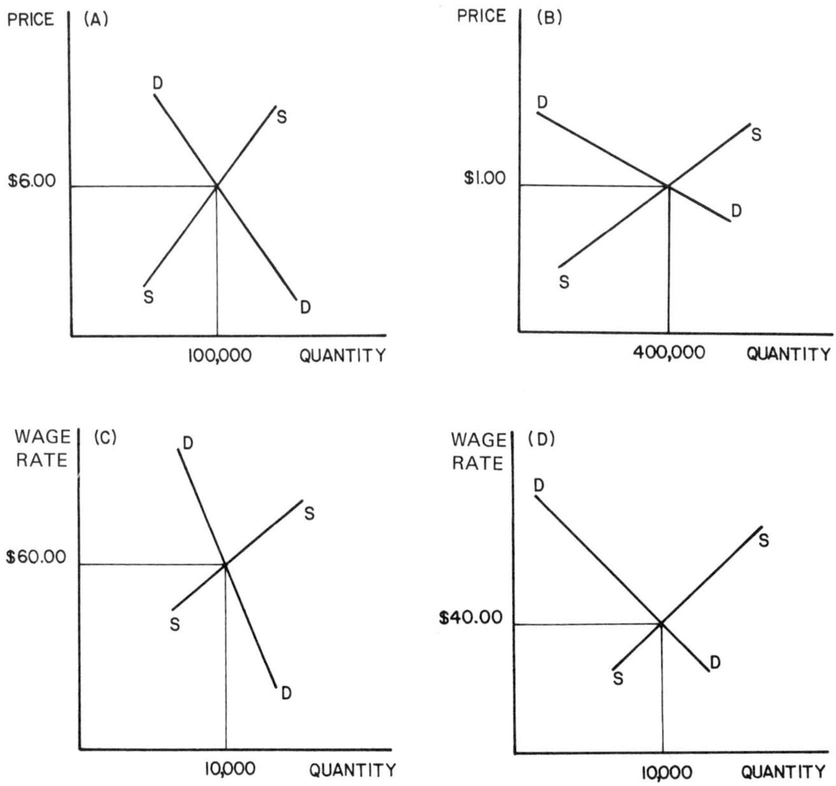

Figure 3.8. Input and output supply and demand--simple economic system. (a) Demand and supply of gasoline. (b) Demand and supply of bread. (c) Demand and supply of refiners. (d) Demand and supply of bakers.

The purchase of the heat pump, however, would tend to shift the demand for such appliances to the right, with resulting changes in the equilibrium price and quantity in this market. Considering the impacts of changes in one market from the perspective of all markets is called general equilibrium analysis. These broader economic impacts must be understood and used in analyzing specific market situations.

A simple general equilibrium model will help us envision cross-relationships. Consider a world in which there are only two products, bread and gasoline, and only two production inputs, bakers and refiners. Assumed initial equilibria in the four markets in this simplified economic system are illustrated in Figure 3.8. Prices are established in each case to clear the markets, and the total income of the bakers and refiners can be calculated by multiplying wage rates by quantity of labor supplied. The total value of each product can be computed by multiplying price and quantity. By definition, the total income to the factors of production must be equal to the total value of the products.

Let us assume that the skills of the baker and the skills of the refiner are identical and that the initial daily wage differential of $20 exists because the bakers don't know what the refiners are earning. If full information becomes available, the bakers will decide to leave the baking industry and shift to refining. The effect is to shift the supply curve for bakers to the left and the supply curve for refiners to the right. The price of labor (wages) will fall in one market and rise in the other. If we assume perfect knowledge and perfect mobility, then workers will move from one occupation to another until wages are equalized--perhaps at $50 per day--as shown in Figures 3.9(c) and 3.9(d). Obviously, repercussions will occur in the product markets. The supply of gasoline will increase and the supply of bread will decrease.* The effects of these shifts will depend both on the magnitude of the shifts and on the price elasticity of demand in each market. Possible new equilibrium conditions in the product markets are shown in Figures 3.9(a) and 3.9(b).

Alfred Marshall, a well-known economist, once drew an analogy between the general equilibrium process and a group of steel balls in a bowl; if you move one of the balls, all the others will shift to new positions. This general equilibrium effect must be recognized when analyzing broad questions of energy policy and when

---

*Quite clearly, it requires more than refiners to produce gasoline. The multiplicity of inputs and outputs in contemporary society makes the general equilibrium process very complex. We have assumed away these complexities in our make-believe world of two factors and two products so that we can concentrate on the equilibrium process.

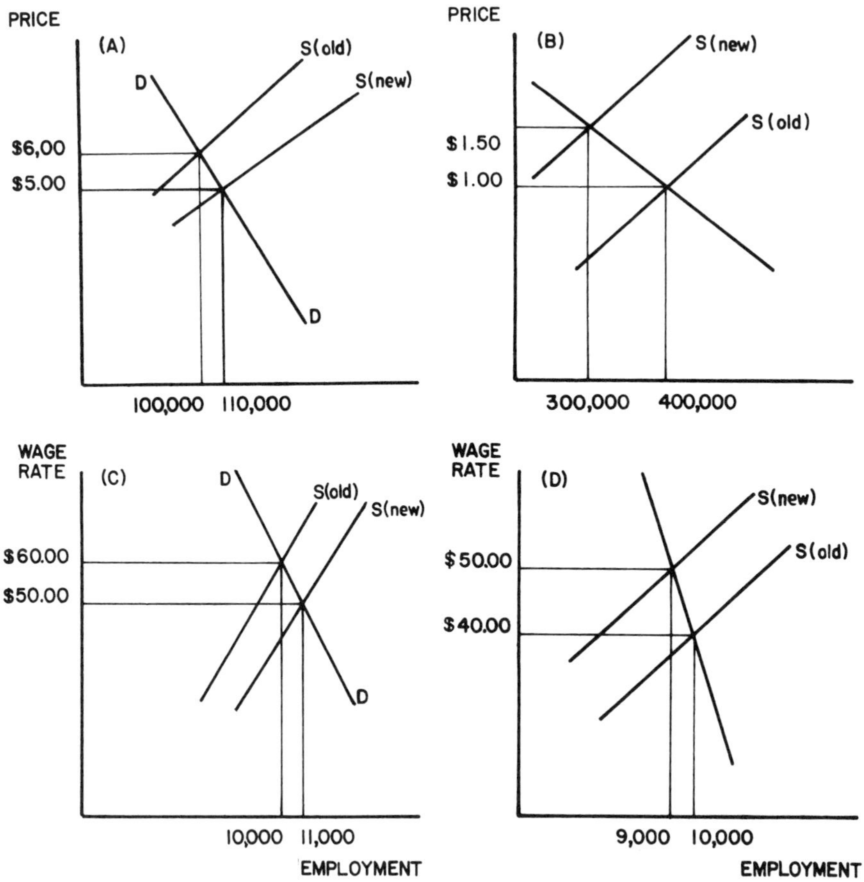

Figure 3.9. Supply and demand in economic system. (a) Demand and supply of gasoline. (b) Demand and supply of bread. (c) Demand and supply of refiners. (d) Demand and supply of bakers.

dealing with the day-to-day decisions involved in managing a business firm or any other organization.

The concepts of demand, supply, market equilibrium, and elasticity discussed in this chapter provide a framework for thinking about broad issues related to energy. The problem situations that follow will test your understanding of these concepts. It usually helps to diagram each situation.

1. Suppose that a tax is levied on the sale of gasoline. The tax is calculated at 50% of the price and is paid by the final consumer to the retailer. How will this affect the equilibrium price of gasoline and the equilibrium quantity? (Assume the initial equilibrium price is $0.50 and the equilibrium quantity is one million gallons.)

2. The same situation exists as in (1), except that the tax imposed is a flat rate of $0.25 per gallon. Will the increase in price be greater than, less than, or the same as in (1)? Explain. What is the new equilibrium quantity? What other differences exist between case (1) and case (2)? (Hint: Consider the effect on government revenues.)

3. As a result of the rapid increase in the price of crude oil in recent years, additional effort has been placed on exploration for oil and several new deposits have been found. In addition, research on techniques of secondary and tertiary recovery has resulted in technical breakthroughs which will eventually lead to greater reserves and lower production costs. Illustrate the effect of these changes on the crude oil market.

4. Increasing consciousness of energy conservation opportunities and tax incentive programs to encourage energy conservation in the home (such as the installation of heat pumps and additional insulation) result in a 10% reduction in electrical energy demand (at all prices). At the same time, new mine safety regulations are expected to add 20% to the cost of mining coal. How will these changes eventually affect the equilibrium price and quantity in the coal market?

SUGGESTED READING

E. F. Brigham and J. L. Pappas, *Managerial Economics,* 2nd ed., Dryden, Hinsdale, Ill., 1976.

R. C. Dorf, *Energy, Resources and Policy,* Addison-Wesley, Reading, Mass., 1978.

P. A. Samuelson, *Economics,* 10th ed., McGraw-Hill, New York, 1976.

Chapter 4

BASIC MANAGEMENT CONCEPTS AND INVESTMENT DECISION CRITERIA

THE PROCESS OF MANAGEMENT

Management is defined as the judicious use of means to achieve ends. It is a process which includes the functions of planning, execution, and control. The purpose of management in any organization is to achieve specific hoped-for end results or goals within definite time spans. The inputs to the process are various types of economic resources such as raw materials, energy, money, equipment, technology, and people. Thus, in sum, management is the conscious, systematic process of utilizing resources in pursuit of preplanned goals. Also, management refers to the people in a particular organization who carry out this process.

The management process is cyclical in nature. Managers perform the primary function of planning when they decide what to do and how to do it. They determine organization or unit goals, resource needs, resource allocations, and specific functional plans. Then they implement or execute these plans after approval by higher level managers or the board of directors.

This execution function involves (1) obtaining resources, (2) organizing, and (3) directing. Human resources, financial resources, materials, equipment, and energy are obtained in sufficient quantity to satisfy operating needs for a future time period--a month, a quarter, a year, or longer. Often contracts are entered into with suppliers to provide energy or other resources for several years in the future. Human resources are organized, or grouped into operating units under specific responsible managers, as specified in the organization plans. Then human endeavors are directed by each manager.

Finally, control, the third basic management function, is exercised. After plans are implemented, resource inputs such as energy, materials, and labor are measured in each time period for each unit of output and actual usage is compared with planned or budgeted usage. If deviations or variances exist, explanations are sought and corrective actions are taken. This process, which starts with the standards or goals set in the planning process and includes measurement, comparison, evaluation, and necessary corrective action, is called managerial control. It is applied to outputs as well as

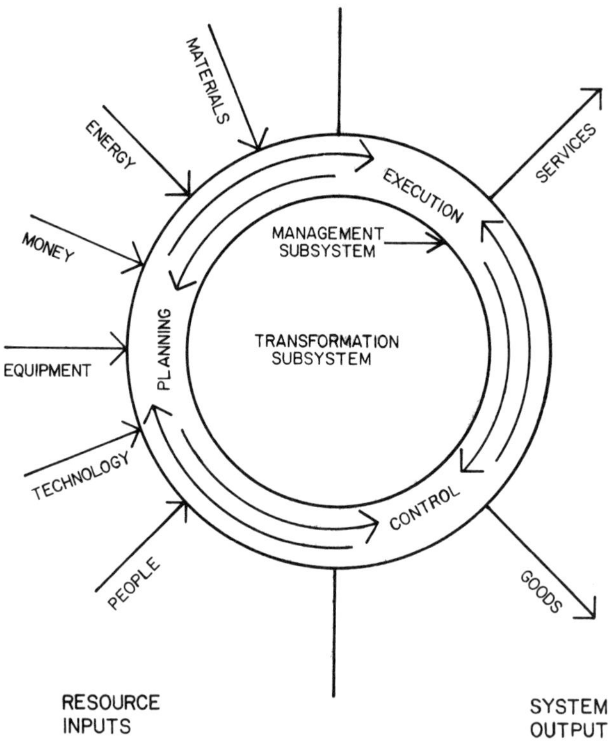

Figure 4.1. Resource flows and the management process in industrial systems.

inputs. For example, important output variables which must be controlled in any industrial firm include product quantity per unit time, product quality, sales per unit time, and profitability per year per dollar of investment.

A simple schematic diagram of an industrial system, including the managerial subsystem, is shown in Figure 4.1. Arrows indicate the flow of resource inputs and salable outputs, as well as the interaction pattern between basic management functions.

BUSINESS OBJECTIVES AND DECISION CRITERIA

Each industrial firm seeks to fulfill a variety of specific objectives or goals. However, practically every firm in a market economy pursues four basic goals:

1. Providing certain goods and services
2. Making a satisfactory level of profits
3. Remaining in business for an indefinite period
4. Fulfilling other broad social responsibilities

The basic role of business firms in a social system is to provide the goods or services demanded by society and to provide them at desired levels of quantity, quality, and cost per unit. Thus, the most fundamental goal of every firm is to fulfill this service role in its selected market niche. At the same time, each firm tries to earn a certain minimum level of profits, which means that total revenues should exceed total expenses by a predetermined amount in a given period. This definition of profit excludes the cost of the equity capital or funds which are provided by stockholders. Thus, one of the basic justifications for profit is to compensate investors for the use of their funds and for assuming the risk of losing them. Profit also provides funds for expansion and improvement of operating facilities within the firm, and in the broad economic subsystem of society, it is the mechanism for allocating scarce resources efficiently.

Profitability goals are stated in many different units:

1. Absolute dollars (related to profits in prior years)
2. Return on stockholders' equity (net worth)
3. Return on permanent capital invested (long-term debt plus stockholders' equity)
4. Return on total assets
5. Return on net sales (revenues)
6. Earnings per share of common stock

Each of these corporation performance indexes may be useful as the basis for setting goals and evaluating results, but most firms use several of them to obtain a more complete picture of performance.

Survival or continuous existence is a goal of most companies, and whenever a company chooses the corporation form of organization, a desire for continuous operation for an indefinite period is implied. To ensure survival a firm must operate within the legal and ethical constraints imposed by society. It must provide society desired products or services. It must earn an acceptable level of profits to ensure continued investor support, and it must manage its cash and other resources so as to remain solvent and be able to operate without major interruptions. For example, failure to pay accounts on time or a disrupted flow of energy could cause shutdowns and eventual bankruptcy of the firm.

In the past decade, more and more firms have broadened their view of social responsibility. Even though they have performed important social roles for many years in providing goods, services, jobs, tax payments, economic stimuli through major purchases, and stabilization within communities, they have been challenged to do more to meet social needs. Many firms have accepted a broader role as corporate citizens and have contributed the time and skills of corporate executives and specialists, monetary resources, and many other types of support in pursuit of worthwhile community and societal goals. Companies have modified their policies, often with a "healthy shove" by the U.S. Congress, to control environmental pollution more effectively, to provide equal employment opportunity for all persons, and, recently, to help conserve available energy resources.

The four basic types of goals are sought simultaneously by industrial firms, even though the profit and service goals are given top priority. Thus, the primary criterion for making decisions on products or services to be provided, on resources to be obtained, on markets to serve, on distribution methods, on pricing, and on hundreds of other issues is the economic one. Before operating changes are made, the expected economic benefits must exceed the cost; before new investments are made, the expected return must be equal to or greater than the firm's cost of obtaining the debt capital and/or equity capital to be invested. However, firms will sometimes take actions to benefit society, even though the actions result in a net cost or negative return, using traditional accounting methods.

INDUSTRIAL SYSTEM COMPONENTS

A system is a group of components which are interrelated and interacting so as to form a unified whole. Every living plant, animal, and person is a system, and these systems are components of broader ecological or social systems. Thousands of inanimate systems exist in the world today which are mechanical, electrical, or electronic in nature. They include vehicles, machines, computers, instruments, household appliances, and buildings.

Every industrial organization is a human system. It is comprised of organizational subunits called divisions, plants, departments, offices, or sections. Each of these operating and support components has a certain function to fulfill, each unit's role is related to that of every other unit, and certain common processes are used. Also, all units are expected to work in harmony to achieve broad organizational goals, even if certain unit goals are not fully realized. In addition to the basic structural components, each individual member, each machine, and each unit of other resource inputs is a component of the total system and of specific internal subsystems. Therefore, when a change is considered for one system component, the impact and the implication for other components must be identified and evaluated before that change is made. For example, when operating changes to conserve energy are considered, the effect of such changes on individuals, departments, other resources, and outputs should be considered. When new investments to conserve energy are proposed, the drain on limited financial resources must be evaluated, as well as the effect on system operation and the relative return on investment.

The basic design of organizations varies widely, and the design will affect the location of energy management programs. Industrial systems are typically organized at the highest level on the basis of either (1) enterprise functions, (2) product lines or type of customer, or (3) geographical territories.

In a functional structure, typical divisions would include production, marketing, finance, research and development, engineering, and personnel management. An effective system for coordinating activities is required with this structure, for no single unit can stand alone. Energy management responsibility will be centralized at the corporate level or assigned to the production manager.

In a product, customer, or geographically structured firm, operating units are more independent and are often designated as "profit centers." In such structures, the manager of each unit will usually be responsible for energy management, but a coordinating energy management function may exist at the corporate level.

The process of dividing organization activities into logical groupings such as those described is called differentiation or departmentation. In order to coordinate the activities of these separate units so that they all behave in the best interest of the single industrial system called the firm, various methods of integration must be created. Mechanisms used for this purpose include committees, coordinators, project or program managers, product managers, and staff specialists with functional authority. Since energy management activities affect many parts of an organization, any of these structural forms may be used to coordinate such efforts.

Another way to classify organizational units is as (1) operating or line units or (2) support or staff units. The operating units are those which contribute directly to fulfilling the primary purpose of an organization. For example, production and marketing units in an industrial firm would be operating units. And, those units which contribute indirectly, such as accounting or maintenance, would be called support units.

When considering energy management programs in industrial systems, it is important to recognize the basic design of the organization and to identify mechanisms for integration. In this way, persons with authority to decide and act can be determined. Also, the impacts and implications of proposed changes in energy utilization can best be recognized if viewed within a known structural design and evaluated for the system as a whole.

OPERATIONS MANAGEMENT AND ENERGY CONSERVATION

The primary concern of an operations manager is to achieve short-term goals as effectively and as efficiently as possible. Effectiveness is the degree to which a goal such as output or sales per time period is achieved, and managerial efficiency is an index of resources required in relation to those actually used. The operations manager will typically establish goals for three important variables:

1. Output and sales of goods or services per time period
2. Quality of the goods or services provided
3. Cost per unit of goods or services

One significant cost element in many operating systems is the energy used in obtaining raw materials, in producing finished goods, in selling products, and in distributing the products to ultimate consumers. Therefore, in order to reduce costs per unit of output and to increase overall efficiency and profitability, every operations manager should seek to reduce energy costs as well as the costs of other inputs. Detailed procedures for better management of energy resources will be discussed in subsequent chapters, but the types of changes which may lead to lower energy costs are

1. Changes in operating procedures or policies which require no capital investment
2. Changes in equipment or facilities which require additional investment
3. Changes in operations which require new knowledge that must be obtained through company research and development programs or from external consultants

The major types of energy costs in industrial systems are incurred for

1. Energy used in production processes
2. Energy used for transporting raw materials, work-in-process, finished goods, and personnel
3. Energy used for space heating, cooling, and lighting, and for the operation of auxiliary equipment

Energy costs incurred by consumers or users of the products manufactured in industrial systems can be reduced by designing more energy-efficient products, but this type of action will be classified and discussed as a strategic rather than an operational change.

When a business firm decides to implement a program to conserve and manage energy, some person or group must be responsible. Several different approaches may be used. First, a corporate executive may be assigned the task of developing, coordinating, and evaluating an energy management program. For example, Olin has designated a corporate Vice President for Planning and Energy. Also, a corporate staff specialist may be designated as the primary energy conservation coordinator in a firm. The Dow Chemical Company has designated a person in this category as Manager, Corporate Energy Conservation.

At the division or plant level the top operating manager may assume the role of energy management program director, or this person may designate a staff or line subordinate to fill this role. Also, committees or task forces may be established at the corporate, division, or plant level to plan and implement an energy management program. External consultants from universities or consulting firms may be employed to assist with these efforts. For such a program to be successful, the key operating managers must support the effort and should become personally involved.

If efforts are not made to manage energy more carefully, externally imposed changes may force such actions through restricted energy supplies, new government regulations, higher energy costs, and failure to compete effectively with more efficient firms. Such efforts may not be justified in firms where energy is a minor cost element, but a firm does not know what the potential for improvement is until actual energy usage, costs, and options are reviewed. In many cases, the benefits of energy management efforts are much greater than the costs, because savings can be realized by a small investment of time in planning and implementing new procedures and policies in operating systems.

LONG-RANGE BUSINESS STRATEGIES TO CONSERVE ENERGY

Much of the discussion of energy conservation in recent years has focused on existing industrial plants, households, commercial buildings, school systems, and other facilities. More attention has been given to short-range, operational changes in existing systems than to long-range, strategic changes in future systems. However, the opportunities to conserve energy will be greater in some firms and other institutions through changes in basic strategies than through modifications in operating systems.

Formal strategic planning has been introduced in many large industrial firms in the past decade. The purpose has been to evaluate investment opportunities and alternative operating policies more systematically in an effort to find the best possible way to use a company's limited resources in pursuit of profits and related goals in a competitive environment. Strategic changes are expected to have a major impact on performance and often alter the basic nature of the firm. New strategies may involve the addition or deletion of product lines as well as changes in markets served. They may include a set of integrated marketing policies aimed at more effective market penetration. They may pertain to vertical or horizontal integration, new sources of supply for materials or energy, changes in facility locations, and new financial arrangements. Also, the introduction of advanced technology in entire production or operating facilities may be a type of strategic plan adopted by firms in the future to conserve energy or to satisfy pollution control regulations.

One type of strategic change which may lead to greater energy conservation among customers is to design and market more energy-efficient products. Also, fuel savings may result from relocation of facilities to get closer to markets or to sources of materials or energy. Investments in entire, new production systems may lead to significant reductions in energy and pollution control costs. Finally, new research and development strategies may lead to more energy-efficient products and processes. Specific examples of such strategic changes by industrial corporations, as well as operational changes, will be discussed in Chapter 7.

SELECTION OF ENERGY-CONSERVING INVESTMENTS

When opportunities to conserve energy are found that require no additional investment, the decision on changing procedures or policies should be relatively easy to make. Only a cost-benefit analysis is required; if the estimated net costs per year to make the change are less than the value of expected net benefits per year, the change should be made. Also, if the projected benefits from proposed research and development are expected to exceed the costs of the effort, after adjustment for the time value of cash inflows and outflows, the research and development program should be implemented. On the other hand, decisions on energy-conserving opportunities which require a capital investment are more difficult and require extensive analysis.

A capital investment is defined as the expenditure of funds for equipment or other items which will provide service for a period of time greater than one year. Examples would include new process equipment, new buildings, insulation, and control instruments. Benefits or cost savings should be realized each year, but the decision to invest or not to invest depends on the initial cost, the value of future cost reductions and cost increases, the timing of cash inflows and outflows, the expected life of the asset purchased, the cost of obtaining capital funds, and the type of budgetary system in effect.*

Several investment analysis methods are used for making such decisions, but only three of the most commonly used ones will be discussed here--the present value method, the rate of return method, and the payback period method.

Present Value Criterion

The present value method converts all projected future cash inflows and outflows related to a proposed investment into units of value at a single point in time, usually the present or the time when the investment will be made. This procedure corrects all monetary flows for the time value of money or the fact that $1000 in hand today is worth more than $1000 which will not be received until next year. Future cash flows are discounted for this loss in value as the waiting period increases, and the procedure is exactly the same as finding the compounded future value of a present pool of cash which is invested at a known rate of interest, but it is in the opposite direction.

When an investment in an energy-saving item such as insulation is considered, the initial cost is recorded, as well as the annual projected savings in energy costs. The useful life of the

---

*With respect to the last variable, some school systems, government agencies, and other institutions separate operating budgets and capital budgets so completely that potential cost savings which would result from a capital investment are given little weight in capital decisions or else the criteria for making new investments are unreasonably difficult to meet (for example, the payback period must be one year or less).

insulation is estimated and the cost of capital* is determined. A discount factor (PV_{sp}), which is used to convert a future cash flow to its present value, is calculated (or obtained from a table+) for each future year of the insulation's life, using the equation:

$$PV_{sp} = \frac{1}{(1 + i)^n}$$

where

PV_{sp} = present fractional value of a future single payment or savings

i = discount rate or cost of capital

n = number of years that the investment will yield benefits

The projected future after-tax savings for each year that the asset can be used are multiplied by the appropriate PV_{sp} factor and the values are summed.† This sum is the net present value of all future cash inflows and outflows, and if it is equal to or greater than the present cost of purchasing the insulation, the investment should be made. Since the decision is based on the net present values of all future cash flows related to the proposed investment, this method of analysis is called the present value criterion. This method is the most precise one available, but it and all others are limited by the accuracy of projected cash inflows and outflows and the value chosen for cost of capital (i).

---

*The cost of capital may be determined from (1) the weighted average cost of long-term debt and equity capital to the company (debt fraction × cost of debt) + (equity fraction × cost of equity) or (2) the highest rate of return the company could realize if the funds were used for alternative investment opportunities.

+A table of present value factors for single payments is given in Table 4.1 and a table of present value factors for annuities is given in Table 4.2.

†If future savings are expected to remain constant each year, the present value of all future savings can be found by multiplying the annual savings by the appropriate PV_a factor found in Table 4.2.

Table 4.1. Present Value Factors: Present Value of One Dollar at the End of n Years

Year (n)	1%	2%	3%	4%	5%	6%	7%	8%	9%	10%	12%	14%	15%	Year (n)
1	0.990	0.980	0.970	0.962	0.952	0.943	0.935	0.926	0.917	0.909	0.893	0.877	0.870	1
2	0.980	0.961	0.943	0.925	0.907	0.890	0.873	0.857	0.842	0.826	0.797	0.769	0.756	2
3	0.971	0.942	0.915	0.889	0.864	0.840	0.816	0.794	0.772	0.751	0.712	0.675	0.658	3
4	0.961	0.924	0.888	0.855	0.823	0.792	0.763	0.735	0.708	0.683	0.636	0.592	0.572	4
5	0.951	0.906	0.863	0.822	0.784	0.747	0.713	0.681	0.650	0.621	0.567	0.519	0.497	5
6	0.942	0.888	0.837	0.790	0.746	0.705	0.666	0.630	0.596	0.564	0.507	0.456	0.432	6
7	0.933	0.871	0.813	0.760	0.711	0.665	0.623	0.583	0.547	0.513	0.452	0.400	0.376	7
8	0.923	0.853	0.789	0.731	0.677	0.627	0.582	0.540	0.502	0.467	0.404	0.351	0.327	8
9	0.914	0.837	0.766	0.703	0.645	0.592	0.544	0.500	0.460	0.424	0.361	0.308	0.284	9
10	0.905	0.820	0.744	0.676	0.614	0.558	0.508	0.463	0.422	0.386	0.322	0.270	0.247	10
11	0.896	0.804	0.722	0.650	0.585	0.527	0.475	0.429	0.388	0.350	0.287	0.237	0.215	11
12	0.887	0.788	0.701	0.625	0.557	0.497	0.444	0.397	0.356	0.319	0.257	0.208	0.187	12
13	0.879	0.773	0.681	0.601	0.530	0.469	0.415	0.368	0.326	0.290	0.229	0.182	0.163	13
14	0.870	0.758	0.661	0.577	0.505	0.442	0.388	0.340	0.299	0.263	0.205	0.160	0.141	14
15	0.861	0.743	0.642	0.555	0.481	0.417	0.362	0.315	0.275	0.239	0.183	0.140	0.123	15
16	0.853	0.728	0.623	0.534	0.458	0.394	0.339	0.299	0.252	0.218	0.163	0.123	0.107	16
17	0.844	0.714	0.605	0.513	0.436	0.371	0.317	0.270	0.231	0.198	0.146	0.108	0.093	17
18	0.836	0.700	0.587	0.494	0.416	0.350	0.296	0.250	0.212	0.180	0.130	0.095	0.081	18
19	0.828	0.686	0.570	0.475	0.396	0.331	0.277	0.232	0.194	0.164	0.116	0.083	0.070	19
20	0.820	0.673	0.554	0.456	0.377	0.312	0.258	0.215	0.178	0.149	0.104	0.073	0.061	20
21	0.811	0.660	0.538	0.439	0.359	0.294	0.242	0.199	0.164	0.135	0.093	0.064	0.053	21
22	0.803	0.647	0.522	0.422	0.342	0.278	0.226	0.184	0.150	0.123	0.083	0.056	0.046	22
23	0.795	0.634	0.507	0.406	0.326	0.262	0.211	0.170	0.138	0.112	0.074	0.049	0.040	23
24	0.788	0.622	0.492	0.390	0.310	0.247	0.197	0.158	0.126	0.102	0.066	0.043	0.035	24
25	0.780	0.610	0.478	0.375	0.295	0.233	0.184	0.146	0.116	0.092	0.059	0.038	0.030	25

Table 4.1. (continued)

Year (n)	16%	18%	20%	22%	24%	25%	26%	28%	30%	35%	40%	50%	Year (n)
1	0.862	0.847	0.833	0.820	0.806	0.800	0.794	0.781	0.769	0.741	0.714	0.667	1
2	0.743	0.718	0.694	0.672	0.650	0.640	0.630	0.610	0.592	0.549	0.510	0.444	2
3	0.641	0.609	0.579	0.551	0.524	0.512	0.500	0.477	0.455	0.406	0.364	0.296	3
4	0.552	0.516	0.482	0.451	0.423	0.410	0.397	0.373	0.350	0.301	0.260	0.198	4
5	0.476	0.437	0.402	0.370	0.341	0.328	0.315	0.291	0.269	0.223	0.186	0.132	5
6	0.410	0.370	0.333	0.303	0.275	0.262	0.250	0.227	0.207	0.165	0.133	0.088	6
7	0.354	0.314	0.279	0.249	0.222	0.210	0.198	0.178	0.159	0.122	0.095	0.059	7
8	0.305	0.266	0.233	0.204	0.179	0.168	0.157	0.139	0.123	0.091	0.068	0.039	8
9	0.263	0.225	0.194	0.167	0.144	0.134	0.125	0.108	0.094	0.067	0.048	0.026	9
10	0.227	0.191	0.162	0.137	0.116	0.107	0.099	0.085	0.073	0.050	0.035	0.017	10
11	0.195	0.162	0.135	0.112	0.094	0.086	0.079	0.066	0.056	0.037	0.025	0.012	11
12	0.168	0.137	0.112	0.092	0.076	0.069	0.062	0.052	0.043	0.027	0.018	0.008	12
13	0.145	0.116	0.093	0.075	0.061	0.055	0.050	0.040	0.033	0.020	0.013	0.005	13
14	0.125	0.099	0.078	0.062	0.049	0.044	0.039	0.032	0.025	0.015	0.009	0.003	14
15	0.108	0.084	0.065	0.051	0.040	0.035	0.031	0.025	0.020	0.011	0.006	0.002	15
16	0.093	0.071	0.054	0.042	0.032	0.028	0.025	0.019	0.015	0.008	0.005	0.002	16
17	0.080	0.060	0.045	0.034	0.026	0.023	0.020	0.015	0.012	0.006	0.003	0.001	17
18	0.069	0.051	0.038	0.028	0.021	0.018	0.016	0.012	0.009	0.005	0.002	0.001	18
19	0.060	0.043	0.031	0.023	0.017	0.014	0.012	0.009	0.007	0.003	0.002		19
20	0.051	0.037	0.026	0.019	0.014	0.012	0.010	0.007	0.005	0.002	0.001		20
21	0.044	0.031	0.022	0.015	0.011	0.009	0.008	0.006	0.004	0.002	0.001		21
22	0.038	0.026	0.018	0.013	0.009	0.007	0.006	0.004	0.003	0.001	0.001		22
23	0.033	0.022	0.015	0.010	0.007	0.006	0.005	0.003	0.002	0.001			23
24	0.028	0.019	0.013	0.008	0.006	0.005	0.004	0.003	0.002	0.001			24
25	0.024	0.016	0.010	0.007	0.005	0.004	0.003	0.002	0.001	0.001			25

From Ezra Solomon, John J. Pringle, *An Introduction to Financial Management*. ©1977, Goodyear Publishing Company.

Table 4.2. Present Value Factors: Present Value of One Dollar per Year for n Years

Year (n)	1%	2%	3%	4%	5%	6%	7%	8%	9%	10%	12%	14%	15%	Year (n)
1	0.990	0.980	0.971	0.962	0.952	0.943	0.935	0.926	0.917	0.909	0.893	0.377	0.870	1
2	1.970	1.942	1.914	1.886	1.859	1.833	1.808	1.783	1.759	1.736	1.690	1.647	1.626	2
3	2.941	2.884	2.829	2.775	2.723	2.673	2.624	2.577	2.531	2.487	2.402	2.322	2.283	3
4	3.902	3.808	3.717	3.630	3.546	3.465	3.387	3.312	3.240	3.170	3.037	2.914	2.855	4
5	4.854	4.713	4.580	4.452	4.330	4.212	4.100	3.993	3.890	3.791	3.605	3.433	3.352	5
6	5.796	5.601	5.417	5.242	5.076	4.917	4.767	4.623	4.486	4.355	4.111	3.889	3.785	6
7	6.728	6.472	6.230	6.002	5.786	5.582	5.389	5.206	5.033	4.868	4.564	4.288	4.160	7
8	7.652	7.325	7.020	6.733	6.463	6.210	5.971	5.747	5.535	5.335	4.968	4.639	4.487	8
9	8.566	8.162	7.786	7.435	7.108	6.802	6.515	6.247	5.985	5.759	5.328	4.946	4.772	9
10	9.471	8.983	8.530	8.111	7.722	7.360	7.024	6.710	6.418	6.145	5.650	5.216	5.019	10
11	10.368	9.787	9.253	8.760	8.306	7.887	7.499	7.139	6.805	6.495	5.938	5.453	5.234	11
12	11.255	10.575	9.954	9.385	8.863	8.384	7.943	7.536	7.161	6.814	6.194	5.660	5.421	12
13	12.134	11.348	10.635	9.986	9.394	8.853	8.358	7.904	7.487	7.103	6.424	5.842	5.583	13
14	13.004	12.106	11.296	10.563	9.899	9.295	8.745	8.244	7.786	7.367	6.628	6.002	5.725	14
15	13.865	12.849	11.938	11.118	10.380	9.712	9.108	8.560	8.061	7.606	6.811	6.142	5.847	15
16	14.718	13.578	12.561	11.652	10.838	10.106	9.447	8.851	8.313	7.824	6.974	6.265	5.954	16
17	15.562	14.292	13.166	12.166	11.274	10.477	9.763	9.122	8.544	8.022	7.120	6.373	6.047	17
18	16.398	14.992	13.753	12.659	11.690	10.828	10.059	9.372	8.756	8.201	7.250	6.467	6.128	18
19	17.226	15.678	14.324	13.134	12.085	11.158	10.336	9.604	8.950	8.365	7.366	6.550	6.198	19
20	18.046	16.351	14.877	13.590	12.462	11.470	10.594	9.818	9.129	8.514	7.469	6.623	6.259	20
21	18.857	17.011	15.415	14.029	12.821	11.764	10.836	10.017	9.292	8.649	7.562	6.687	6.313	21
22	19.661	17.658	15.937	14.451	13.163	12.042	11.061	10.201	9.442	8.772	7.645	6.743	6.359	22
23	20.456	18.292	16.444	14.857	13.489	12.303	11.272	10.371	9.580	8.883	7.718	6.792	6.399	23
24	21.244	18.914	16.936	15.247	13.799	12.550	11.469	10.529	9.707	8.985	7.784	6.835	6.434	24
25	22.023	19.523	17.413	15.622	14.094	12.783	11.654	10.675	9.823	9.077	7.843	6.873	6.464	25

Table 4.2. (continued)

Year (n)	16%	18%	20%	22%	24%	25%	26%	28%	30%	35%	40%	50%	Year (n)
1	0.862	0.848	0.833	0.820	0.807	0.800	0.794	0.781	0.769	0.741	0.714	0.667	1
2	1.605	1.566	1.528	1.492	1.457	1.440	1.424	1.392	1.361	1.289	1.225	1.111	2
3	2.246	2.174	2.107	2.042	1.981	1.952	1.923	1.868	1.816	1.696	1.589	1.407	3
4	2.798	2.690	2.589	2.494	2.404	2.362	2.320	2.241	2.166	1.997	1.849	1.605	4
5	3.274	3.127	2.991	2.864	2.745	2.689	2.635	2.532	2.436	2.220	2.935	1.737	5
6	3.685	3.498	3.326	3.167	3.021	2.951	2.885	2.759	2.643	2.385	2.168	1.824	6
7	4.039	3.812	3.605	3.416	3.242	3.161	3.083	2.937	2.802	2.508	2.263	1.883	7
8	4.344	4.078	3.837	3.619	3.421	3.329	3.241	3.076	2.925	2.598	2.331	1.922	8
9	4.607	4.303	4.031	3.786	3.566	3.463	3.366	3.184	3.019	2.665	2.379	1.948	9
10	4.833	4.494	4.193	3.923	3.682	3.571	3.465	3.269	3.092	2.715	2.414	1.965	10
11	5.029	4.656	4.327	4.035	3.776	3.656	3.544	3.335	3.147	2.752	2.438	1.977	11
12	5.197	4.793	4.439	4.127	3.851	3.725	3.606	3.387	3.190	2.779	2.456	1.985	12
13	5.342	4.910	4.533	4.203	3.912	3.780	3.656	3.427	3.223	2.799	2.469	1.990	13
14	5.468	5.008	4.611	4.265	3.962	3.824	3.695	3.459	3.249	2.814	2.478	1.993	14
15	5.576	5.092	4.676	4.315	4.001	3.859	3.726	3.483	3.268	2.826	2.484	1.995	15
16	5.669	5.162	4.730	4.357	4.033	3.887	3.751	3.503	3.283	2.834	2.489	1.997	16
17	5.749	5.222	4.775	4.391	4.059	3.910	3.771	3.518	3.295	2.840	2.492	1.998	17
18	5.818	5.273	4.812	4.419	4.080	3.928	3.786	3.529	3.304	2.844	2.494	1.999	18
19	5.878	5.316	4.844	4.442	4.097	3.942	3.799	3.539	3.311	2.848	2.496	1.999	19
20	5.929	5.353	4.870	4.460	4.110	3.954	3.808	3.546	3.316	2.850	2.497	1.999	20
21	5.973	5.384	4.891	4.476	4.121	3.963	3.816	3.551	3.320	2.852	2.498	2.000	21
22	6.011	5.410	4.909	4.488	4.130	3.971	3.822	3.556	3.323	2.853	2.499	2.000	22
23	6.044	5.432	4.925	4.499	4.137	3.976	3.827	3.559	3.325	2.854	2.499	2.000	23
24	6.073	5.451	4.937	4.507	4.143	3.981	3.831	3.562	3.327	2.855	2.499	2.000	24
25	6.097	5.467	4.948	4.514	4.147	3.985	3.834	3.564	3.329	2.856	2.499	2.000	25

From Ezra Solomon, John J. Pringle, *An Introduction to Financial Management*. ©1977, Goodyear Publishing Company.

Rate of Return Criterion

The rate of return method is a simple accounting procedure which does not correct for the time value of money, but if future annual cash savings are expected to remain constant, it will provide a sound basis for investment decisions.

The following equation can be used to calculate the average percentage return (R) on the average investment over the life of the asset:

$$R = \frac{100(S - D - I - \Delta T)}{\bar{I}}$$

where

S = net monetary operating savings per year

D = depreciation expense per year

I = average interest expense per year on the average investment

ΔT = incremental tax expense on (S - D - I)

$\bar{I}$ = average investment, or the average of the initial investment plus the ending investment, assuming funds are borrowed and repaid in equal installments annually

A minimum return for making investments should be established, based on the cost of capital or the return available from alternative uses of funds. In many firms, a 10 to 20% return must be projected before an investment will be made.

Payback Period Criterion

The payback period method determines the time required to recover the initial cash investment through net operating savings each year. The payback period (P) is calculated as follows:

$$P = \frac{I}{S - \Delta T}$$

where

I = initial investment

S = annual operating savings

ΔT = incremental tax expense on the annual operating savings

This method does not reflect the time value of money or differences in the useful lives of competing proposed investments. Therefore, differences in the time patterns of future cash flows and in projected asset lives must be considered. For example, if two investment alternatives have similar projected payback periods, the one with the earliest cash inflows or the longest expected life should be selected.

In sum, the present value criterion is the most useful one for evaluating investment opportunities, but the rate of return method provides useful supplementary data for direct comparison with actual companywide returns on equity and assets and the payback method gives a measure of the cash recovery time. Investment proposals for energy-saving projects must be carefully evaluated, as capital is also a scarce resource and should be used wisely in order to attract funds from investors in the future. An illustration is provided in Chapter 6 for analyzing an energy-saving investment opportunity using the three criteria described here.

SUGGESTED READING

J. Argenti, *Systematic Corporate Planning,* Halsted, New York, 1974.

D. I. Cleland and W. R. King, *Systems Analysis and Project Management,* 2nd ed., McGraw-Hill, New York, 1975.

C. W. Hofer and D. Schendel, *Strategy Formulation: Analytical Concepts,* West Publishing, St. Paul, Minn., 1978.

J. G. Longnecker, *Principles of Management and Organizational Behavior,* 4th ed., Charles E. Merrill, Columbus, Ohio, 1977.

E. Solomon and J. J. Pringle, *An Introduction to Financial Management,* Goodyear, Santa Monica, Calif., 1977.

Part 2

ENERGY MANAGEMENT IN OPERATING FACILITIES

In order to utilize energy resources in the most efficient manner in any operating system, some degree of formality in fulfilling the energy management function is considered essential. Granted, each manager is responsible for utilizing each resource (including energy) as efficiently as possible, but energy-related decisions often go beyond the scope of the job of any single manager. Thus, to give adequate attention to efficient energy utilization in a period when energy is becoming much more expensive and supplies more uncertain, and to deal with systemwide energy actions, a separate function to coordinate energy management activities in a given organization is desirable.

In Chapter 5, practical considerations for planning, implementing, and controlling a formal energy management program are discussed. Categories of energy conservation opportunities are reviewed, and selected analytical and operating procedures for more efficient energy utilization are described.

An energy audit is essential in any successful energy management program. Methods for making and using such an audit (including a billing and a field audit) are explained in Chapter 6. Also specific steps for analyzing each energy conservation opportunity are outlined and illustrated in this chapter.

In Chapter 7, three case studies are presented to indicate exactly how energy management can be practiced in various settings. Case 1 is a study of an engineering office building, Case 2 deals with a mining equipment manufacturer, and Case 3 pertains to a glass-decorating plant.

Chapter 5

ENERGY MANAGEMENT PROGRAMS

THE MANAGEMENT OF ENERGY

The concepts and principles of management were discussed in a previous chapter. Energy is one of the resources subject to the management process and this has led to the widespread use of the term "energy management." It will become evident in the following discussion that energy conservation and energy management are closely related concepts, although energy management has a broader connotation than energy conservation. Schipper [1] has given a definition of energy conservation, which is sufficiently broad to serve as a definition of energy management, as follows: "The strategy of adjusting and optimizing energy-using systems and procedures so as to reduce energy requirements per unit of output (or 'well-being') while holding constant or reducing total costs of providing the output from these systems."

Historical Factors in Energy Management

It is often said that energy management is an idea whose time has come, and it is useful to consider briefly some of the factors which have prevented the concept of energy management from developing earlier on a broad scale in industrial and commercial firms. It should be noted in passing, however, that in some energy-intensive industries the concept of energy management is well established.* The main reason that energy management has not received more widespread special attention until recent years (i.e., the post-1973 oil embargo period) has been the relatively low cost of energy in the past. Coupled with the low cost was the widely held perception that our energy resources were virtually limitless, and very little attention was devoted to the problem of energy shortages. There has also been, until recently, an absence of information and understanding of the potential benefits to be derived from energy management coupled with a tendency to equate energy conservation with deprivation and a reduced standard of living. A typical viewpoint, expressed as recently as 1972, is contained in a publication of the Chase Manhatten Bank [2], which states:

> Analysis of the uses of energy reveals little scope for major reduction without harm to the nation's economy and its standard of living. The great bulk of the energy is utilized for essential purposes -- as much as two thirds is for business related reasons. And most of the remaining third serves essential private needs. Conceivably, the use of energy for such recreational purposes as vacation travel and the viewing of television might be reduced -- but not without widespread economic and political repercussions. There are some minor uses of energy that could be regarded as strictly non-essential -- but their elimination would not permit any significant savings.

During the past four or five years, however, a broad base of opinion and supporting documentation [3-12] has emerged to indicate that a large potential for energy conservation exists through effective energy management.

ELEMENTS OF AN ENERGY MANAGEMENT PROGRAM

Recognizing energy as a resource to be managed in parallel with land, labor, capital, and raw materials suggests that the key elements of energy management will be the same as the elements involved in the management of other resources. The several elements that must be present in a successful energy management program are grouped under

*For example, the Du Pont Corporation formally recognized the need for energy management in 1903 by establishing the Heat, Light and Power Group within the corporate Engineering Department.

the three basic functions of management discussed in Chapter 4: planning, execution, and control.

Planning

The planning process begins with the realization that energy is a resource to be managed and the decision to see that adequate management is indeed applied. It continues, in due time, with the establishment of realistic goals. Programs necessary to achieve the goals should be defined, and procedures should be established for systematic monitoring of results. This can only be done after information on energy flows and cost has been obtained. Energy management requires successive phases of planning, execution, and control, so that information obtained in each phase provides a basis for planning the next one.

Execution

The execution of the program involves, most importantly, a commitment on the part of the management within the company. This commitment must flow from the highest level of management and must include a commitment of the necessary personnel, time, and funding to carry out a successful program. Program responsibility should be assigned to one individual, who should be given an appropriate title such as Energy Management Coordinator and who should be located on an organization chart in a manner which shows without ambiguity his or her direct access to the top management person within the firm. The role of the Energy Management Coordinator is to develop and implement an energy management program capable of achieving prescribed energy use goals. Depending on the size of the firm, it may be appropriate to establish an Energy Management Committee to assist the Energy Management Coordinator. It is essential, however, that accountability and responsibility for the program be vested in one individual and not a committee. It may also be desirable and necessary to acquire the services of an outside professionally trained energy management consultant to assist the Energy Management Coordinator. These decisions, made in the planning phase, are implemented when people are assigned specific energy management functions.

After the organizational phase of an energy management program is completed, the energy data base must be established and it should include both internal and external energy information. The internal data base refers to knowledge about the use of energy within the firm whereas the external data base includes knowledge about energy use in general, current and future availability and costs of various forms of energy, and available energy conservation technology.

A central component of the information gathering process is the energy audit, which is discussed in detail in Chapter 6. It involves collecting data on historical total energy use from such sources as utility bills and information on current in-plant use from energy use surveys. The survey part of the audit process is essential in order to observe energy-consuming processes and facilities and to identify potential energy conservation opportunities.

In the next phase, each potential energy conservation opportunity should be analyzed with respect to the capital expenditure needed to implement the action and the annual energy cost savings at projected energy costs, and the cost of capital or required return on investment. Then, plans for taking advantage of each selected energy conservation opportunity should be developed and implemented. They may involve changed procedures, new capital equipment, and even research or development.

Control

During each phase of an energy management program, goals are established and action plans are developed and implemented. Then actual results should be measured and evaluated by comparing them to the goals, and corrective action should be taken. These efforts to see that desired results are achieved constitute managerial control.

Specific control functions such as monitoring and reporting may be delegated by the Energy Management Coordinator, while the evaluation and correction functions are retained. To aid in the monitoring process, submeters can be installed on large energy-consuming devices or groups of devices.

The interaction of the three components of management is evident as the implementation of the program proceeds and results are evaluated; plans for saving energy are often modified, new goals may be set, and additional monitoring may be required. A successful energy management program requires continual planning, execution, and control.

IMPLEMENTING AN ENERGY MANAGEMENT PROGRAM

Top Management Commitment

The implementation of an energy management program begins with the decision to see that energy is managed. It must be led by an obvious commitment on the part of the management. The person identified as having the responsibility for the program must be assured of sufficient resources to formulate and implement the details of the program. With this commitment, planning for the next phase of the program can start. In due course management may be required to commit further resources for the implementation of specific projects or action-proposals resulting from the planning effort.

The top levels of management should inform supervisors of the economic reasons for the need to conserve energy and their responsibility for implementing energy-saving actions in the areas of their accountability. They should be provided with guidelines as to what is expected of them. The guidelines should cover such actions as:

1. Developing ideas and plans for enlisting employee support and participation
2. Planning and participating in energy audits
3. Surveying the literature on ways to conserve energy and communicating these ideas and suggestions
4. Establishing realistic and achievable energy conservation goals
5. Developing uniform record keeping, reporting, and energy accounting
6. Planning and conducting a continuing program of activities to stimulate interest in energy conservation efforts

Goals should be set for energy savings. A preliminary goal should be established at the start of the program with provision for a revised goal based on savings potential estimated from the results of the audits. Management should communicate periodically to employees to reinforce the emphasis on energy conservation and to report on progress.

Establishing a Data Base

The necessary data base can be established by an audit and survey of energy use and related activities within the facility. The data base must be sufficiently comprehensive to establish clearly the historical trends in energy use as well as the current energy situation. It is from this reference point that success of the energy management program will be judged. The data base must also identify the forms of energy used, their relative importance to the company's continued success, and the distribution of energy end use. Such information is essential in formulating responses to mandatory fuel cutbacks or serious changes in availability and cost of fuels. Yet another important use of the data base is in the evaluation of potential energy conservation opportunities (ECOs). Such an evaluation is aimed at determining the potential benefits from any energy-related action and frequently requires a detailed knowledge of current energy use, either in the facility as a whole or in a particular process or piece of equipment. The energy saving to be expected from any action taken is an essential factor in subsequent investment decisions and a reliable data base provides the necessary information from which to estimate such savings.

An adequate data base cannot be established by a single isolated act. Rather, it is the result of a concerted first effort in several directions and the refinements of continual probing and investigation. The initial thrust may be accomplished by a thorough study of fuel billing data and production records and a detailed survey of the facility. The value and significance of the results of these activities are sufficient to warrant a separate chapter of this book and the reader is referred to Chapter 6 for further insights into the gathering of bulk energy data.

This data gathering will continue as time proceeds. In addition, the need for special surveys will arise. Detailed mapping of steam, gas, compressed air, or electrical systems may become necessary. Surveys of a facility at special times, such as weekends or nights, may be appropriate. Surveys should be made to determine where additional energy metering is needed. Such metering may be essential for the fine-tuning of an energy management project or required for follow-up monitoring. An incidental benefit of submetering may be the psychological or competitive effect it can have on operators associated with submetered equipment.

Taking Advantage of Energy Conservation Opportunities

The energy management program continues to take advantage of ECOs. The identification of ECOs, begun as a part of the data gathering process, must of necessity be pursued with diligence on a continuing basis. That new ECOs will continue to arise is an inevitable result of the continually changing schedule within the plant in response to economic and technical developments and the gradual degradation of equipment performance with time and use. The response to a particular ECO will depend on the specific nature of the case, but generally one of two courses is open.

Where energy waste can be eliminated by simple maintenance or rescheduling actions involving no cost, such actions should be taken immediately. Such maintenance is of the type that can be done by in-house staff.

Where energy conservation opportunities involve more effort than is available from routine maintenance and where equipment purchase is necessary, a more complete study is required. From the list of energy conservation projects evolving from the audits and surveys made to assemble the data base, projects should be evaluated according to the following procedure.

1. Estimate the annual energy savings for each project.
2. Project future energy costs using available future energy cost projections and estimate the annual dollar savings.
3. Estimate the capital cost for each project.
4. Evaluate the investment merit of each project.
5. Assign priorities to projects on the basis of investment merit.
6. Select conservation projects for implementation and request capital spending authorization.
7. Implement authorized projects.

In this way a logical and orderly approach to improved energy management can be followed, thus permitting optimum use to be made of funds as they become available. Clearly, the estimates and evaluations will have to be updated from time to time, and continued effort on the part of the Energy Management Coordinator will bring to light other projects which will need to be evaluated and inserted at the appropriate level in the prioritized list.

Finally, the experience gained in retrofit action in existing facilities should be put to use by reviewing the design of all capital projects such as new plants, expansions, buildings, and equipment, to assure that efficient utilization of energy is incorporated in the designs. In these cases consideration must also be given to the question of future availability of energy in general and the various fuel forms in particular.

Monitoring and Follow-up

The control phase of the management process is, in the case of energy management, as difficult as it is important. Since measurement is the prerequisite of control, the monitoring of energy savings becomes an essential part of the energy management program. Energy savings must ultimately be measured as reductions in energy purchases, which reinforces the need for a good data base. But it is a complex problem to determine with any degree of certainty the extent of the energy saved. Efforts to evaluate the savings of individual projects can be and are used to justify capital investments. Ultimately, however, the savings must be seen in reduction of energy use, otherwise they cannot be considered real savings.

An important reason for insisting on seeing reductions "on the meter" is that in complex systems it is easy to save in one place at the expense of energy use somewhere else. The systems approach to energy conservation is an important concept, and is no more so than in the monitoring of energy use to evaluate energy savings. The main problem, other than the fact that an effort to conserve energy in a particular part of a facility may be offset by a corresponding increase elsewhere, is that energy consumption and production data contain large levels of "noise"--random effects which mask the real situation. Some of the noise can be filtered out, as explained in the next chapter, and some can be explained by observation of the process; work stoppages, inferior-quality raw materials, bad weather, and significant national or international events can all cause temporary but drastic changes in production activity and energy efficiency. Even when these are considered, there still will be some unaccounted for variations in energy use. Clearly, any conservation action, or group of actions, must produce an effect greater than this random noise or it will be insignificant, undetectable, and therefore not justifiable.

The measurement of energy savings by monitoring energy consumption after an energy management program has been implemented is closely associated with the data gathering that has been stressed as an important preliminary step in the process. Now the energy consumption following conservation efforts can be compared with energy consumption before those efforts began. In only a few cases will it be possible to compare the periodic (daily, monthly, or yearly) energy consumptions directly. Even in facilities such as office and commercial buildings, where the use is constant, there will be variations in energy consumption due to varying weather conditions which must be accounted for in making before/after comparisons. In manufacturing facilities where a single product is made, it may be fairly easy to tie energy consumption to production level. The concept of energy used per unit of output is easily understood in such a situation, and is very useful in monitoring energy savings.

However, for many facilities there may be more than one product and the mix of products may very well vary over even short periods of time. In such cases it may be possible, by using the results of a complete energy audit or by statistical analysis of combined historical energy and production records, to divide the total energy use among the various product lines. Otherwise, some other measure of activity within the facility, such as payroll dollars or gross sales, must be used. This constitutes a normalization of energy use. A plot against time of normalized energy use can give a much clearer picture of energy savings than can be obtained by looking at energy consumption figures alone. A documented example is shown in Figure 5.1 for boiler fuel used in the processing of textile goods. It is not possible to give specific instructions on how to carry out this normalization because of the wide range of circumstances which surround energy use. Each Energy Management Coordinator is encouraged to search out the best measure of energy use for the specific situation. In some cases industry standards may develop which have the additional advantage of permitting comparison

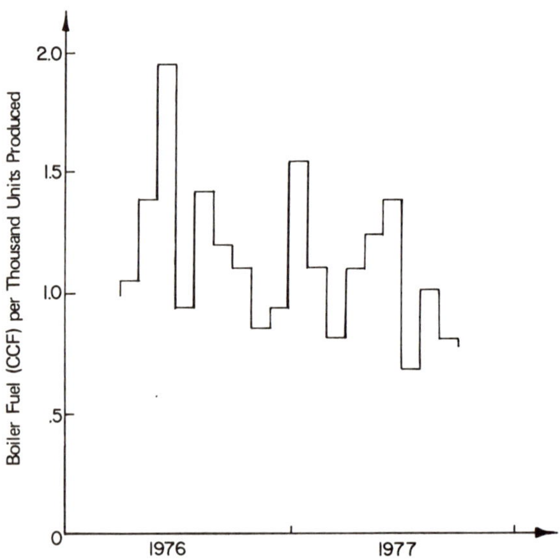

Figure 5.1. Fuel consumption per unit of goods produced.

of energy efficiency between plants and companies. In whatever way it is done, this monitoring provides a useful measure of current performance, and when monitoring is continuous, it can point up the success of the program and areas where further work is necessary. Thus, the monitoring phase of the program provides documented confirmation of energy savings which, in turn, provides greater leverage for the Energy Management Coordinator to implement further projects.

CATEGORIES OF ENERGY CONSERVATION OPPORTUNITIES

A key element of the energy management process is the identification and analysis of energy conservation opportunities. They can be classified in the following three categories.

Procedural and Maintenance ECOs

Procedural and maintenance ECOs are those which can be accomplished at little or no cost by changing procedures and improving maintenance of energy-consuming devices. Examples of this type of ECO include turning off lights and equipment when not needed, reducing lighting to levels recommended for the particular tasks being performed, and changing thermostat settings to levels appropriate for comfort (for example, 68°F in the winter and 78°F in the summer when the building is occupied).

A procedural action of importance which can lead to significant energy cost savings but not necessarily a reduction in energy consumption in the facility involves an analysis of demand charge for various energy forms, particularly electricity. The utility bills should be examined to determine if significant demand charges are being incurred. If it is found that demand charges are being incurred, the daily energy use cycle should be analyzed on an hourly basis to determine those portions of the load which are contributing to the peak demand. An hourly profile of energy consumption can be established by reading the utility meters hourly, subtracting successive hourly readings to obtain the hourly consumption, and then plotting hourly consumption in the form of a histogram over a 24-hour period. Care should be exercised to insure that the 24-hour period being examined represents a typical day of operation. A load profile which displays hourly peaks should be analyzed with a view toward obtaining a flatter profile either by the use of demand limiters (manual or automatic) or by shifting peak loads to other times during the day where possible to give a flatter load profile. In facilities that involve more than one-shift operations, load leveling can sometimes be achieved by moving certain energy-consuming operations which contribute to peak loads to another shift. The local utility is a good source of technical assistance in analyzing and correcting a load demand problem.

Modifications Using Available Technology

After the procedural and maintenance ECOs have been identified and implemented, the next level of ECOs to be considered involve the application of existing "off the shelf" technology requiring capital expenditures. Competent technical expertise (engineering and/or architectural) is required in this process to analyze the technical feasibility of making equipment and/or process changes, and to develop the data on which to base an investment analysis of the proposed project. If the firm does not have the necessary in-house technical expertise required for this analysis, outside assistance should be obtained from a reputable engineering consulting firm. Examples of this category of ECO include the addition of insulation, waste heat recovery, improved controls on combustion equipment, improved controls on high-voltage ac equipment, and computerized energy management systems.

ECOs that Require Research and Development

A final category of ECO involves longer range planning requiring research and development activities. This category of ECO cannot be implemented by the application of existing technology. An example of such an ECO would be a totally new process change which would produce a given level of production output with a lesser expenditure of energy.

TOOLS AND CONCEPTS APPLICABLE TO ENERGY MANAGEMENT

It is clear from the definition and discussion of the elements of an energy management program that energy management is a broader concept than energy conservation. The ultimate objectives of a successful energy management program are to provide the same level of well-being or productivity with a lower expenditure of energy and assure an adequate supply of needed energy at the lowest possible cost to carry out the firm's mission. There are a number of management tools and concepts that will be useful in carrying out a successful program.

Energy Life Cycle Costing

Purchase of energy-using equipment or facilities requires consideration of two important aspects: the initial cost of the equipment and the running costs. The lowest initial cost may be associated with equipment that is characterized by inefficient use of energy and whose total cost, over its useful life, thus will prove greater than that which would have been incurred by the purchase of a more expensive but more energy-efficient device. The calculation of the total cost of purchase and lifetime operation of the equipment is referred to as life cycle costing. While energy efficiency frequently is bought by high first cost, and simply adding more insulation to a hot water storage tank is a good example, this is not universally true. As technological innovation is applied to the energy efficiency problem, both first cost and running costs may be reduced and continual attention to lifelong cost of energy-using plant is essential.

Computerized Energy Management Systems

In facilities where there exist extensive, complex, or numerous energy systems (particularly heating, ventilating, and air-conditioning systems), it becomes very difficult to manage the energy use efficiently in a manual mode. In multiuse facilities, such as department stores and large commercial buildings, the interactions of different systems require continuous monitoring. Where there are many widespread systems, as on a college campus or in a large hotel, manual control becomes impossible because of the time and distance involved. In such situations computerized management systems are used. Sensors throughout the system monitor the current status of the variables being managed and the information is relayed to a central computer. The energy control strategies applicable to the particular situation are implemented by the computer software, and activators throughout the system respond through such actions as lowering thermostat settings, opening or closing dampers in ventilation systems, shutting off or starting up motors driving fans, pumps, or compressors, and extinguishing lights in unused areas. The same approach is also used in complex, energy-intensive industrial processes, and may be tied in with product mix, production rate, and quality control.

The management strategies of computer-controlled systems vary from simple control of electrical peak demand to sophisticated programs which continually optimize the energy effectiveness of a plant or process.

Computer Simulation

As a result of an increasing interest in and concern for effective energy management, many new products and concepts are being promoted. Where the capital outlay for the purchase of new products or the implementation of new ideas is extensive and their ultimate or relative effectiveness is in doubt, it is desirable to have a reliable method of predicting the outcome. Such prediction is possible in some cases--for example, in building design--through computer simulation programs.

Simulation of thermal performance of a large building permits design, equipment, and material options to be evaluated prior to final construction by comparing simulated building energy use for the various options. Rehabilitation of older buildings to an improved energy efficiency can also be accomplished by using simulation studies. The larger programs are versatile and detailed; they require information about the construction and equipment in every corner of the building, as well as the hourly details of the proposed operating schedule and projected annual hour-by-hour data on ambient weather conditions. Such studies require considerable computer run time, and therefore cost, but their benefits are great and lasting. The payback period was short in one project where computer simulation was used to analyze building energy use [13]. A detailed account of the study is contained in Case 1 of Chapter 7.

Energy Cascading

In light of the second law of thermodynamics, it is apparent that some energy is better than other energy. In fact, energy at high temperature, or fuel with high temperature potential, is capable of being used in a wider range of applications than energy at a lower temperature. The idea of energy cascading is to put this fact to use in designing a process in which energy is used through a series of processes, each at a lower temperature than the last. Careful process design and, frequently, the use of heat recovery equipment are required in implementing this concept and some additional energy may be needed to transport the energy to the next stage, but the overall savings can be dramatic. An example of the process would be the generation of high pressure steam in a boiler, its use to provide mechanical power through a pressure reducing steam turbine, the use of the exhaust steam to meet a process steam requirement, and perhaps the recovery of heat from the hot final product for use in preheating hot water or space heating. Two stages of energy cascading are common throughout the industrial sector in preheaters where in early stages of a process, boiler feedwater, combustion air, or makeup air may be heated by some means from an exhaust stream at a higher temperature than that which can be usefully employed at the second stage. Higher levels of cascading require greater innovation, and perhaps more sophisticated equipment, but the rewards in terms of energy saved will be proportionately greater also.

Equipment Efficiency

The question of efficiency has been touched upon in discussing life cycle costing. The potential efficiency of energy-using equipment is a function of design and construction. There are many situations in which this efficiency may not be realized as a result of performance degradation. The incorrect ratio of fuel to combustion air in a boiler, the fouling of boiler tubes, and the restriction of air flow due to dirty filters are all cases where correct maintenance procedures could substantially improve energy use efficiency.

A further net efficiency increase may frequently be realized through the addition of heat recovery equipment to extract normally wasted heat from exhaust gas streams such as flue stacks and ventilators in kilns and drying ovens. These devices do not in themselves change the efficiency of the original equipment but do improve the overall energy efficiency of a plant.

The design efficiency of equipment is an important matter in considering energy management. It is being improved in many areas through technical developments [14] and performance standards [15]. Improved efficiency permits the same level of output for a reduced energy consumption and is, therefore, very much a component in Schipper's concept of energy conservation.

REFERENCES

1. L. Schipper, Raising the productivity of energy utilization, *Annual Review of Energy*, Vol. 1, Annual Reviews Inc., 1976.

2. *Energy Outlook in the United States to 1985*, Chase Manhattan Bank, New York, 1972.

3. *Potential for Energy Conservation*, Office of Emergency Preparedness, 1972, U.S. Government Printing Office, Washington, D.C.

4. *Energy Conservation Potentials*, Shell Oil Co., Houston, 1973.

5. E. Kovach, ed., *Technology of Efficient Energy Utilization*, Scientific Affairs Division, NATO, Brussels, 1974.

6. D. Large, *Hidden Waste*, Conservation Foundation, Washington, D.C., 1973.

7. *Conservation and Efficient Use of Energy*, Pts. 1-4, U.S. House of Representatives, Subcommittee on Science and Astronautics, May 1974, U.S. Government Printing Office, Washington, D.C.

8. *Energy Conservation and S 2176*, U.S. Senate, Committee on Interior and Insular Affairs, Aug. 1973, U.S. Government Printing Office, Washington, D.C.

9. *Energy Waste and Energy Efficiency in Industrial and Commercial Activities*, U.S. Senate, Committee on Commerce, May, June 1974. Also *Industry Efforts in Energy Conservation*, U.S. Senate, Committee on Commerce, Oct. 1974, U.S. Government Printing Office, Washington, D.C.

10. Ford Foundation, Energy Policy Project, *A Time to Choose: America's Energy Future*, Ballinger, Cambridge, Mass., 1974.

11. L. Schipper, *Energy Conservation: Its Nature, Hidden Benefits, Hidden Barriers*, Rept. No. LBL 3295, Lawrence Berkeley Lab., Berkeley, Calif., 1975.

12. L. Schipper, *Efficient Energy Use*, Rept. No. ERG-75-08, Energy and Resources Group, University of California, Berkeley, 1975.

13. E. E. Choat, J. H. Corley, T. C. Fisher, and J. M. Morrison, An evaluation of energy alternatives in an engineering office building, *ASHRAE TRANSACTIONS*, Vol. 82, Pt. 2, 1976.

14. C. Starr, Technical innovations: The answer to resource depletion, *Proceedings of the International Conference on Energy Use Management*, Vol. III/IV, Pergamon, New York, 1977.

15. ASHRAE Standard 90-75, American Society of Heating, Refrigerating, and Air-Conditioning Engineering, Inc., New York, 1975.

Chapter 6
MAKING AND USING AN ENERGY AUDIT

INTRODUCTION

The availability of a reliable energy information data base has already been emphasized as an important component of an energy management program. This data base must be as accurate, complete, and current as possible. It is frequently nonexistent at the time an energy management program is set up, so an energy audit is made to start building such a data base. This chapter discusses the practical aspects of the energy audit with respect to what it is, how it is conducted, and how the results can be used.

THE ENERGY AUDIT

The basic concept of an energy audit is quite simple. It involves an analysis of a facility to determine the forms of energy used, the quantities and costs of the various forms of energy used, the purposes for which the energy is being used, and the identification of energy conservation opportunities. The energy audit is the most important element of a successful energy management project since it provides the starting point and direction of progress for the whole program.

The audit process consists of two principal phases: (1) the billing audit, and (2) the field audit. Each of these phases will be discussed separately.

The Billing Audit

The first phase of the audit is data collection and analysis based on available energy consumption and cost records, as well as on production records in manufacturing facilities. The purposes of this phase of the audit process are

1. To examine historical energy consumption, energy costs, and production levels for trends and anomalies
2. To determine the approximate amount of energy used for space conditioning (weather related) and for production processes (approximately fixed)
3. To determine energy consumption per unit of production, or an equivalent, normalized energy consumption index

The principal source of information concerning historical energy consumption and cost is the utility bill. In spite of the apparent simplicity of such bills, experience has shown that many energy consumers do not understand all of the information they contain and pay little attention to any of the information except the dollar cost. In addition to the information presented on utility bills, the consumer should also be thoroughly familiar with the rate structure. It includes features such as the base charge, relation between consumption level and unit charge, demand charge formula, and a fuel adjustment clause (where applicable).

The current applicable rate should be obtained from the utility and the monthly bills checked. In particular it should be noted that most electrical rates vary in blocks depending on the amount of energy used. This becomes important in evaluating the cost of energy to be saved since a reduction in consumption will mean a reduction of energy at a block rate which is probably not the average rate. Understanding the monthly bill will help to make it clear from whence the savings will come and will permit a more accurate calculation of dollar savings. If assistance is needed in interpreting utility bills, it can be obtained from the customer's utility company.

Examples of forms which may be useful in tabulating production and energy consumption data are illustrated to show the detail

Area of Facility [ ] ft2

Date [ ]

Location [ ]

Volume of Facility [ ] ft3

Month	Year	No. of Units Produced	Operating Cost Level	Operating Level	Other Meas. of Output	Comments
January						
February						
March						
April						
May						
June						
July						
August						
September						
October						
November						
December						

Figure 6.1. Monthly production record form.

Data by [ ]

Date [ ]

Location [ ]

Month	Year	Demand KW	Energy KWh	Demand Charge—$	Energy Charge-$	Total Cost-$
January						
February						
March						
April						
May						
June						
July						
August						
September						
October						
November						
December						

Figure 6.2. Monthly consumption of electricity.

required in this phase of the data gathering process. The form in Figure 6.1 may be used for tabulating monthly production data. Those in Figures 6.2 through 6.5 may be used for recording monthly energy consumption of electricity, natural gas, propane, and fuel oil.

Energy quantities for various types of fuel must be expressed in a consistent set of units for comparison. Conversion factors to obtain Btu equivalents for this purpose were given in Chapter 2. The information that should be extracted from utility bills includes:

1. Total energy consumed (kWh/month or therms/month)
2. Energy cost
3. Rate of energy use (demand, kW)
4. Demand charge
5. The billing period

It is desirable to obtain these data for a minimum period of two years and to plot them for each calendar month in the form of a bar graph or histogram. The billing period shown on the bill must be

Data by ______ Date ______

Location ______

Heating Value ______ Btu/ft^3

Month	Year	Quantity Thousand c.f.	Demand Charge-$	Energy Charge-$	Total Cost-$	Comments
January						
February						
March						
April						
May						
June						
July						
August						
September						
October						
November						
December						

Figure 6.3. Monthly consumption of natural gas.

Data by ____ Date ____

Location ____

Heating Value ____ BTu/gal

Month	Year	Quantity Gallons	Demand* Charge-$	Energy** Charge-$	Total Charge-$	Comments
January						
February						
March						
April						
May						
June						
July						
August						
September						
October						
November						
December						

Figure 6.4. Monthly consumption of propane.

**may be left open if zero*

***may be left open if same as total change*

Data by ____ Date ____

Location ____

Heating Value ____

Month	Year	Quantity Gallons	Total Cost
January			
February			
March			
April			
May			
June			
July			
August			
September			
October			
November			
December			

Figure 6.5. Monthly consumption of fuel oil.

noted, since it rarely corresponds to the calendar month. One method for expressing consumption on a calendar month basis is to use a linear interpolation of consumption for the preceding and following months. The following interpolation formula may be used for this purpose:

$$E_n = \frac{C_n B_n}{B_n + N_{n-1} - B_{n-1}} + \frac{C_{n+1}(N_n - B_n)}{B_{n+1} + N_n - B_n}$$

where

E_n = energy consumption for month n

N_n = number of days in month n

N_{n-1} = number of days in previous month n - 1

C_n = energy consumption for billing period between date B_{n-1} of month n - 1 and date B_n of month n

C_{n+1} = energy consumption for billing period between date B_n of month n and date B_{n+1} of month n + 1

An example where this formula is used is shown in the following table of billing data based on three consecutive billing periods. The irregularity of the billing periods is typical of that encountered in practice.

Billing period	Energy consumption, kWh
12/04/76-1/20/77	23,700
1/20/77-2/06/77	19,000
2/06/77-3/05/77	20,500

It is desired to find the energy consumption for the calendar month of February 1977. In the notation of the foregoing formula, month n is February, month n - 1 is January, and month n + 1 is March. The following values of the parameters are to be used in the formula:

$B_{n-1} = 20$ $\quad B_n = 6$ $\quad B_{n+1} = 5$

$C_n = 19{,}000$ $\quad C_{n+1} = 20{,}500$

$N_{n-1} = 31$ $\quad N_n = 28$

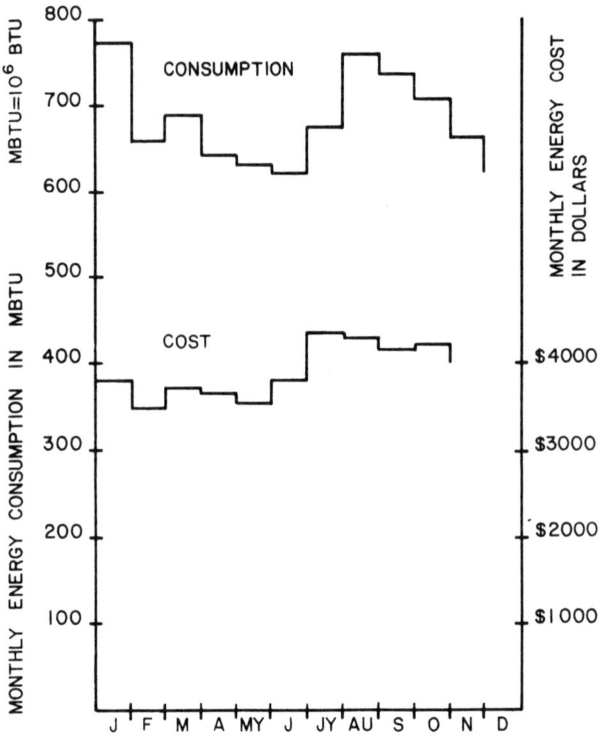

Figure 6.6. Electrical energy consumption, year of 1976.

Solving for E_n gives E_n = 23409 kWh as the energy consumption for February. If the value of 19,000 kWh contained on the bill received in February had been used, an error of more than 18% would have resulted.

An example of an energy-use histogram is shown in Figure 6.6. The advantage of plotting rather than tabulating the data is that trends and variations are perhaps more readily apparent by inspection. After plotting historical energy consumption and cost data for the previous two-year period, the month utility data should be plotted on a continuing basis. Such a continuous plotting procedure will readily indicate trends and sudden shifts in the energy consumption and costs and will serve as a useful tool in the evaluation of the energy management program.

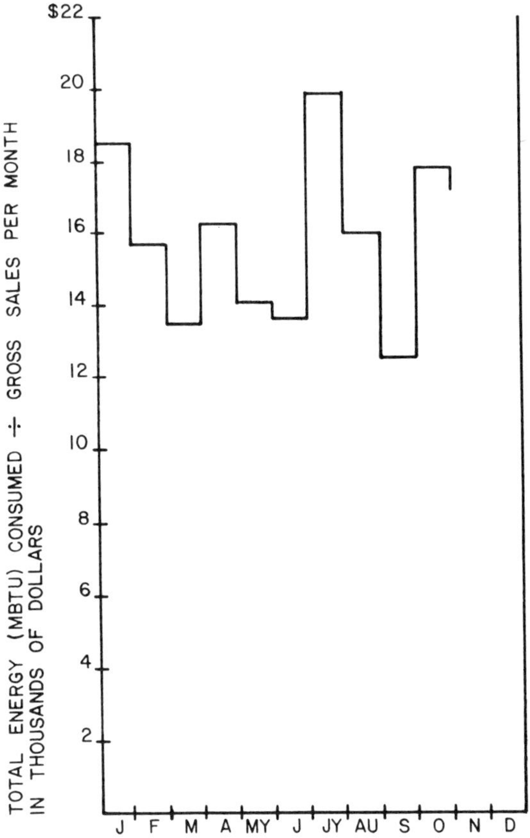

Figure 6.7. Total energy consumption per monthly gross sales in thousands of dollars.

Production data should be obtained and plotted as a histogram for the same period for which energy consumption is plotted. Output should be expressed in units used by the firm such as quantity/week, gross sales, cost of goods sold, dollar value added, raw material costs, or other units appropriate to the normal accounting practices of the firm. A suitable production parameter should be sought with which to monitor on a monthly basis the ratio of total energy consumption to total production. Figure 6.7 shows a histogram of the ratio of total energy consumed to total production in which gross dollar sales are the parameter used to measure production.

Information obtained from the billing audit will not be useful generally in measuring energy usage by function. However, an approximate allocation can be made between space heating energy and process energy by recognizing that process energy use will remain approximately constant (with constant production) and space heating energy will be season-dependent. To separate these components, the monthly energy consumption values for the five months May through September should be taken. The highest and lowest consumption months from the group should be excluded and consumption for the three remaining months should be averaged. This average value can be multiplied by 12 to obtain a reasonable estimate of the annual process energy used, provided the monthly level of production activity is reasonably constant during the year. Upon subtracting the annual process energy determined in this manner from the total annual energy use, a reasonable estimate of the energy used for space heating is obtained. This procedure has proved valid within about 10% accuracy, provided there are no special summer loads, such as air conditioning, which are not associated with production.

The Field Audit

After completing the billing audit phase, the next phase is the field audit or the "facility walk-through audit." The purposes of this phase of the audit process are

1. To allocate energy use by function, physical location, department, or other appropriate division
2. To observe the operation of processes and facilities from an energy-use perspective
3. To identify potential energy conservation opportunities

The field audit involves gathering information about every energy-consuming device in the facility. The ideal information to have includes the operating schedule and rate of energy consumption for each device, but, in practice, one must generally settle for something less complete.

Prior to walking through the facility, it is worthwhile to study whatever drawings of the facility and installed equipment that may be available. But although much useful information can be obtained from drawings, a word of caution needs to be added concerning their reliability. It is not uncommon to find that changes have been made in a facility that were not shown on the drawings. This fact must be kept in mind when attempting to obtain energy information from drawings.

One may question the necessity of a facility walk-through to acquire equipment power ratings in facilities where accurate records of equipment schedules are maintained and facility drawings are available. But even if such information is provided, and is complete, there is still a great value to be derived from walking through a facility and observing first hand what is going on with respect to energy consumption.

One type of information to be obtained from the walk-through audit is energy-related data from nameplates for each energy-consuming device in the facility. A form which is suitable for recording the nameplate data is shown in Figure 6.8. From the information recorded on such audit sheets, energy consumption may be determined by function, physical location, department, or other appropriate divisions. The calculation of energy consumption is based on the fact that energy used is the product of power drawn and the time period of interest (Chapter 2). The column entries in Figure 6.8 permit a determination of these quantities on a monthly basis for each piece of equipment considered. The first column is to record the name plate power rating of the device. The next column is to record the number of units, in case multiple identical units are installed. The next two columns are to record data concerning the operating schedule of the device to permit an estimate of the number of hours per month the device is operated. The column labeled "percent utilization" is to record an estimate of the average loading of the device over the time noted in columns 2 and 3, expressed as a percentage of rated full load. This is sometimes a difficult quantity to estimate, and the experience of the equipment operator is generally helpful in obtaining this information. The column labeled "conversion factor" may be used to convert the energy use

Date [ ]

Data by [ ] Location [ ]

Monthly Consumption Measured in [ ]

Rating	Units	Hr/Day	Days/Wk	Hr/Mo	%Util'n	Conv. Fact.	Energy Used	Devise Name	Monthly Variations ?

Figure 6.8. Data sheet for walk-through energy audit.

as determined from the units obtained from the nameplate data to the energy units indicated in the upper right-hand corner. For example, if it were desired to convert energy in kWh to Btu, the conversion factor would be 3412. The next column is available to enter the calculated monthly energy consumption in the indicated units and the last two columns are reserved for identification and location of the device. Some facilities have major equipment items identified by code numbers. Similarly, it may be advantageous to establish location codes before beginning the audit in order to simplify the data logging process. Having completed such a form at the site (columns 1-4, 6, 9, 10) the energy consumption for all items of interest can be calculated. By using the methods of Chapter 2, the energy used per week or month can be estimated by multiplying the rate at which energy is used, or the power of the equipment by the time the equipment is used. Care must be given to the units in which the power rating and running time are measured since these

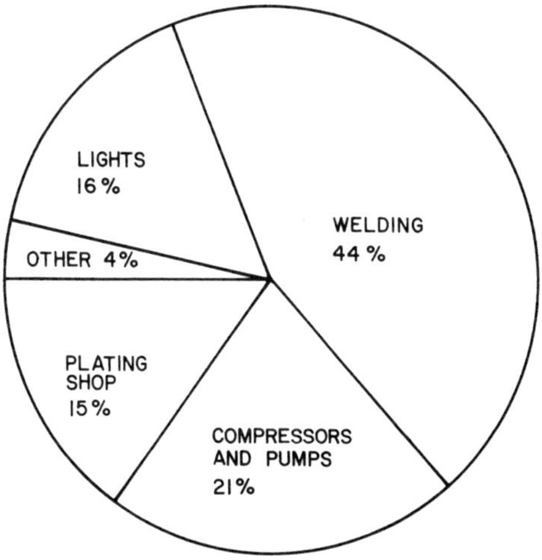

Figure 6.9. Division of electrical energy end use.

will determine the units in which the energy used is calculated. The units (therms, kWh, million Btu, for example) should be chosen to suit the particular facility being studied, but more important, one set of units must be used consistently throughout a single study.

A useful way of displaying the energy consumption data is by means of a "pie diagram" as shown in Figure 6.9. From the "pie diagram" one may determine the areas of greatest energy consumption and, therefore, the areas on which to focus attention in the system modification stage of the energy management program.

In addition to obtaining information on energy consumption by equipment, devices, and lighting, the walk-through audit provides an opportunity to observe, first hand, potential energy conservation opportunities. At this point a checklist becomes a useful reminder of opportunities to be kept in mind. Such checklists are not, however, a license for the uninformed and unobservant to become skilled auditors. Identification of potential ECOs frequently calls for a sophisticated technical background in a very specialized area. At all times, awareness of basic engineering principles and mental alertness are required. Notes should be made during the walk-through so that each potential ECO can be analyzed for its energy-use savings potential and energy-cost savings potential.

ANALYSIS OF ENERGY CONSERVATION OPPORTUNITIES

What may be considered the "bottom line" of the energy audit process is the identification and analysis of potential ECOs. The different types of ECO are discussed in Chapter 5. Whatever its category, each ECO should be analyzed independently following a standardized format. A brief analysis of each ECO should be written up in a form which can be easily understood by the individual with the ultimate authority for deciding whether or not to implement the ECO.

A convenient format for writing up each ECO would include the following items:

1. Identification of the ECO
2. Description of current practice
3. Observations
4. Proposed action
5. Anticipated savings
6. Investment analysis
7. Recommendations

Further examples of ECO analyses can be found in the complete audit reports contained as case studies in Chapter 7.

Example of an ECO Analysis

The subject of this analysis is the reduction of heat loss from the open surface of a heated liquid. The analysis will follow the steps outlined in the foregoing.

1. *Identification of ECO.* Reduction of heat loss from open surfaces of plating tanks.

2. *Current practice.* Three tanks in the electroplating area are heated with hot water which is generated by the gas-fired boiler. Each of the tanks is 15 feet in diameter and 15 feet deep, and open on the top surface. The exposed surface area of heated

electrolyte solution is 177 ft^2 per tank. The temperatures maintained in the three tanks are 130 ± 10°F, 160 ± 5°F, and 185 ± 5°F, respectively.

3. *Observations*. Polypropylene balls had been used to insulate the tank surfaces but were abandoned because of the inconvenience of having some of the balls scattered around the floor when workpieces were removed from the tank. However, the effectiveness of polypropylene balls in reducing heat loss from a heated free surface has been demonstrated and reported in the literature.

4. *Proposed action*. It is recommended that 1-in.-diameter polypropylene balls again be used on the surfaces of the three heated tanks.

5. *Anticipated savings*. Assume a surface heat transfer coefficient of 20 Btu/hr-ft^2-°F [1]. The heat loss from the three tanks assuming ambient temperature of 80°F is

$$Q = (20\ \text{Btu/hr-ft}^2\text{-°F})(177\ \text{ft}^2)[(130 - 80)\text{°F} + (160 - 80)\text{°F} + (185 - 80)\text{°F}] = 0.83 \times 10^6\ \text{Btu/hr}$$

Assuming three-shift operation, seven days per week, with 5% downtime for maintenance gives an annual heat loss from the plating tanks of 6.89×10^9 Btu/yr. Trade literature of polypropylene ball suppliers indicates that heat loss can be reduced by 60 to 90%. Assuming a 75% reduction in heat loss gives an annual savings of

$$(0.75)(6.89 \times 10^9\ \text{Btu/yr}) = 5.17 \times 10^9\ \text{Btu/yr} = 5.17 \times 10^4\ \text{thm/yr}$$

The heat for the plating tanks is supplied by a gas-fired boiler. Assuming an overall efficiency of 75% means that input energy from the natural gas must be

$$(1/0.75)(5.17 \times 10^4\ \text{thm/yr}) = 6.89 \times 10^4\ \text{thm/yr}$$

Assuming the gas to cost \$0.20/thm gives an annual cost savings of

$$(\$0.20/\text{thm})(6.89 \times 10^4\ \text{thm/yr}) = \$13{,}780/\text{yr}$$

Trade literature from a supplier of 1-in.-diameter polypropylene balls indicates that 144 balls/ft^2 of surface should be used.

The total number of balls required plus 10% for loss would be

$$(1.1)(144)(3)(177) = 84{,}110 \text{ balls}$$

The quoted price for this quantity is $27 per thousand, so the total cost will be $2271. Assuming additional labor of one hour per shift at $5.00 per hour to pick up balls from the floor gives a total annual labor cost of $5475 for maintaining the balls. Thus, the net annual before-tax savings will be the difference between fuel cost savings and increased labor costs, or

$$\$13{,}780 - \$5475 = \$8305/\text{yr}$$

If the income tax rate is 48%, the after-tax savings will be $8305 (1 - .48), or $4318.60.

6. *Investment analysis.* The investment analyses made here are based on the three methods discussed in Chapter 4 in order to illustrate the application of those methods.

If the payback criterion is used to evaluate the proposed capital investment in polypropylene balls, the payback period is calculated and compared with the company's desired payback period. The actual payback period for the proposal would be

$$2271 \div 4318.60 = 0.53 \text{ yr}$$

Since this period is just over six months, the investment would be attractive in any organization unless capital funds were extremely scarce or costly.

If the return on investment criteron is applied, the calculated value of the average return on the average investment is

$$\frac{100(8305 - 830.50 - 124.91 - 3527.80)}{\frac{2271 + 227.10}{2}} = \frac{3821.79(100)}{1249.05} \quad \text{or} \quad 306\%$$

assuming a 10-year life for the polypropylene balls, no salvage value, straight-line depreciation, an interest rate of 10% on the average investment, and an income tax rate on incremental savings of 48%. When a return on investment of 306% is projected, this is an extremely attractive opportunity, as the average return on stockholders' equity in major U.S. corporations usually lies in the 12 to 16% range.

If the present value investment criterion is used to evaluate the polypropylene ball proposal, the net cash outflow of $2271 is compared with the present value of net cash inflow (or savings) for 10 years, which is

$4318.60 \times 6.145^*$ or $26,537.80

7. *Recommendation.* Since the cash inflow exceeds the cash outflow by a significant amount, the investment is, by any criterion, attractive. Thus, the proposed action is an investment which is very desirable and should be implemented.

REFERENCE

W. M. Rohsenow and H. Y. Choi, *Heat, Mass, and Momentum Transfer*, Prentice-Hall, Englewood Cliffs, N.J., 1961, p. 102

SUGGESTED READING

S. L. Baron, *Manual of Energy Saving in Existing Buildings and Plants*, Vols. 1 and 2, Prentice-Hall, Englewood Cliffs, N.J., 1978.

J. C. Dale, Energy audit--First step in an energy conservation program, *SME Technical Paper*, Series EM 76-104.

F. S. Dubin and C. G. Long, Jr., *Energy Conservation Standards*, McGraw-Hill, New York, 1978.

F. S. Dubin, H. L. Mindell, and S. Bloome, *How to Save Energy and Cut Costs in Existing Industrial and Commercial Buildings*, Noyes Data Corp., Park Ridge, N.J., 1976.

D. A. Reay, *Industrial Energy Conservation*, 2nd ed., Pergamon, Oxford, England, 1979.

J. M. Shirley, Industrial energy management--First steps are easy, *Industrial Engineering*, Vol. 10, No. 5, May 1978.

W. L. Stebbins, How a big user of energy manages to cut waste, *Power Transmission Design*, Vol. 20, No. 1, Jan. 1978.

A. Thurman, *Plant Engineers and Managers Guide to Energy Conservation*, Van Nostrand Reinhold, New York, 1977.

G. Vanderweil, Savings begin with an energy audit, *Consulting Engineer*, Vol. 47, No. 5, Nov. 1976.

---

*This is the value from Table 4.2 of PVa for an annuity which continues for 10 years at a discount rate of 10%.

Chapter 7

ENERGY MANAGEMENT CASE STUDIES

The practice of energy management is a very specific activity and cannot be dealt with completely in general terms. The strategy developed for any particular situation will depend in part on the specific details of that situation and in part on the specific interests, objectives, and ingenuity of the manager developing the strategy. Three case studies are presented here, not as examples of what must be done, but rather as illustrations of what can be done. They also serve to support points made in the text and to show how techniques discussed there might be implemented. The case studies appear essentially as they were originally written, and while the styles differ widely, some underlying common areas are readily apparent--particularly the need for a quantified data base, the identification and evaluation of energy conservation opportunities, and some form of investment analysis.

The use of computer modeling as a means of evaluating a number of alternatives for energy conservation in a building is illustrated in Case 1, a report [1] by E. E. Choat, Superintendent of Environmental Control Engineering (ECE), Engineering Division, Oak Ridge National Laboratory (ORNL), Oak Ridge, TN; J. H. Corley, Design Supervisor of ECE at ORNL; T. C. Fisher, Design Engineer, ECE at ORNL; and J. M. Morrison, Designer, ECE at ORNL. Cases 2 and 3 have been selected from a number of studies carried out for small industries by groups at the University of Tennessee. Case 2 is particularly interesting as it comprises the final report of a team of graduate students--Winston Redford, Ralph D. Chambers, and William W. Pflager--in a course being taught at the University of Tennessee from the material contained in this book. Case 3 is a report of a study made by two of the authors (William T. Snyder and Frederick W. Symonds) and is included to illustrate how a simplified format can be developed to permit comprehensive free-standing reports to be written for each of a series of industrial studies.

CASE 1: AN EVALUATION OF ENERGY CONSERVATION ALTERNATIVES IN AN ENGINEERING OFFICE BUILDING*

Building 1000 is a 25-year-old office building, located at Oak Ridge National Laboratory, Oak Ridge, TN. It is a two-story, wooden frame structure which has been recently covered with aluminum siding. Most of the building's 48,000 ft^2 (4460 m^2) of floor area is utilized as offices for approximately 200 engineering personnel. Other characteristics of the building which were considered in evaluating energy conservation alternatives are as follows:

1. Walls and floors--wood, with no insulation
2. Roof--flat, built-up, with ceiling and 3 inches (7.6 cm) of blown wool insulation
3. Windows--double-hung, with aluminum frames
4. Lights--2465 fluorescent bulbs, 40 watts each
5. Ventilation--4700 cfm (133 cmm) exhaust for toilets and blueprint machines
6. Occupancy--8:00 a.m. to 4:30 p.m., weekdays
7. Heating system--electric baseboard heaters
8. Cooling system--160 package air conditioning units
9. Equipment--miscellaneous small equipment, such as copiers, vending machines, water coolers, and blueprint machines

Building 1000 has no characteristics which would make it particularly wasteful or conservative of energy. The building is seen as representative of any number of office buildings which were built prior to the oil embargo of 1973. Therefore, it provides an

*From *ASHRAE TRANSACTIONS*, Vol. 82, Part 2, copyright 1976. Used by permission of American Society of Heating, Refrigerating and Air-Conditioning Engineering, Inc.

opportunity to evaluate energy conservation alternatives which exist in buildings constructed prior to the energy conservation era and which are common throughout the country.

Method of Evaluation

To determine the effect of various alternatives, yearly energy requirements were established for a base case and nine alternative modes of building operation. Base case operating conditions are those which existed prior to the energy conservation era. Alternatives were evaluated by changing one parameter of the case previously studied and recalculating yearly energy requirements. The effect of the alternative in question was then established by the differences in energy required. Alternatives considered were

1. Modified lighting schedule
2. Reduced lighting levels
3. Modified ventilation schedule
4. Equipment shutdown
5. Additional floor insulation
6. Storm windows
7. Reset inside temperatures
8. Daylight savings time--April to October
9. Daylight savings time--year around

Calculations were made with a computer which was programmed to consider

1. Average hourly weather data
2. Thermal properties of walls, roof, and windows
3. Area and orientation of structure surfaces
4. Hourly occupancy, lighting, and equipment profiles
5. Internal equipment and lighting loads
6. Thermostat settings
7. Type of heating and cooling systems

The computer used the energy analysis code originally developed for the U.S. Postal Services. Various simulation options were added to the program to better suit the needs of this study. These revisions included seasonal temperature reset, fan system shutdown, and a subroutine for simulating the performance of the packaged air conditioning units.

Evaluations and Results

Base Case

The operating conditions assumed for this case were typical of those which existed prior to the energy conservation era. The building was assumed to be operating as follows: (1) five days per week with 8:00 a.m. to 4:30 p.m. occupancy, (2) full lighting on a continuous basis, and (3) heating and air conditioning as required to maintain 75°F (23.9°C) year-round. On this basis the annual energy consumption was found to be 2,413,600 kWh per year. At today's cost of electricity* in Oak Ridge, this represents a yearly cost of $36,500. The distribution of energy to lighting, heating, and cooling is shown in Table 7.1.

To account for the 3 to 1 cooling/heating ratio, it should be noted that under these conditions the building balance point was found to be approximately 40°F (4.4°C). That is, at outside temperatures above 40°F (4.4°C), cooling was required; below 40°F (4.4°C), heating was required. High lighting loads, which constituted 50% of the energy used, contributed significantly to this condition.

Modified Lighting Schedule

For this case, it was assumed that lights would be turned off by janitors when their shift was completed at 11:00 p.m. and that lights would remain off on weekends. For the purpose of computation, it was assumed that lighting would be reduced to 10% of full load during these off periods.

Energy required after turning lights off according to the described schedule is identified in Case 1 of Table 7.1. As seen, significant reductions in energy required for lighting and cooling are realized, but the energy required for heating increased. The increase in heating energy is required to offset heat from lights which is unavailable during off periods.

---

*Power costs as of 8-1-75: $15.16 per 1000 kWh.

Table 7.1. Energy Consumption--Building 1000*

Alternatives	Lighting		Heating		Cooling		Total electrical		
		%Δ		%Δ		%Δ		%Δ	Σ %Δ
Base case	1215.7	-	275.9	-	921.9	-	2413.6	-	-
Case 1	720.9	-40.7	553.5	+100.6	585.8	-36.5	1860.2	-22.9	-22.9
Case 2	526.4	-16.0	628.2	+27.1	434.3	-16.4	1588.9	-11.2	-34.1
Case 3	526.4	-	387.0	-87.4	414.9	-2.1	1328.3	-10.8	-44.9
Case 4	526.4	-	105.4	-102.0	210.6	-22.1	842.4	-20.1	-65.0
Case 5	526.4	-	81.6	-8.6	218.7	+0.9	826.7	-0.7	-65.7
Case 6	526.4	-	63.1	-6.7	221.8	+0.3	811.2	-0.6	-66.4
Case 7	526.4	-	43.7	-7.0	215.9	-0.6	785.9	-1.0	-67.4
Case 8	526.4	-	47.9	+2.0	204.4	-1.2	778.7	-0.3	-67.7
Case 9	526.4	-	57.7	+3.5	199.5	-0.5	783.5	+0.2	-67.5

Case 1: Lighting schedule change
Case 2: Reduce lighting levels
Case 3: Night shutdown of exhaust air
Case 4: Night shutdown of A/C units
Case 5: Add first floor carpet
Case 6: Add storm windows
Case 7: Reset temperature, 78°/68°
Case 8: Daylight savings time, April-October
Case 9: Daylight savings time, year around

*1000's kWh

Savings in excess of that indicated are now being realized by turning lights off on an office-by-office basis as clean-up is completed.

Reduced Lighting Levels

For this case, it was assumed that lighting levels would be reduced to levels more consistent with recently published guidelines. For the purpose of computation, lighting loads were reduced from 2.5 W/ft^2 to 1.65 W/ft^2 (26.9 to 17.8 W/m^2). Other conditions were assumed the same as the previous case.

As seen in Table 7.1, Case 2, reduced lighting loads resulted in a further decrease in the energy required for lighting and cooling and a further increase in energy required for heating.

Savings approximately equal to that indicated are being realized. Light bulbs have been physically removed from selected fixtures. The removal of these was directed by an electrical engineer who measured and adjusted lighting levels in accordance with current guidelines for general and task lighting. A total of 1024 40-watt bulbs were removed.

Reduced Ventilation

For this case, it was assumed that mechanical ventilation systems would be shut down during unoccupied periods, and would run only from 7:00 a.m. to 5:00 p.m. on weekdays. These systems (a total of 4700 cfm) (133 cmm) consist of toilet and reproduction area exhaust fans.

As seen in Table 7.1, Case 3, heating energy usage was reduced significantly while cooling energy showed only a 2% drop. A relatively small savings in cooling energy is reasonable, since summer nights in eastern Tennessee are generally cool and little cooling of ventilation air would be required.

Savings approximately equivalent to the amount indicated are being realized. Automatic timers have been installed to operate building exhaust fans on the schedule described.

Equipment Shutdown

For this case, it was assumed the heating and air conditioning equipment would be shut down during unoccupied periods. During shutdown periods, the building temperature would be allowed to seek an equilibrium condition. As seen in Table 7.1, Case 4, significant savings of energy for both heating and cooling are attainable provided that unacceptable building temperatures are not experienced during shutdown periods.

To better understand the behavior of building temperatures under shutdown conditions and to make an assessment of the feasibility of shutting down equipment, temperatures were monitored over a period of time when the equipment was shut down. Temperatures at five building locations, along with outside temperature, were recorded on a continuous basis. An example of the data recorded is shown in Figure 7.1.

As seen, all building temperatures decreased and tended to follow outside temperatures when heating equipment was shut down. As would be expected, floor surface temperatures and temperatures near outside walls were most affected. Initially temperatures at thermostat level decreased sharply but then leveled out somewhat while outside temperature continued to decrease. The significant difference between temperatures at thermostat level and floor level accounts for the numerous complaints of cold floors and also raises the question of an appropriate setting for such a space. Clearly, a setting of 68°F (20°C) would result in floor temperatures of approximately 58°F (14.4°C) and an unacceptable condition within the space.

From these data it was concluded that (1) equilibrium temperatures during night shutdown conditions would not be so low as to cause serious discomfort, and (2) thermostats should be set higher than 68°F (20°C) in order to maintain comfort conditions at the floor level.

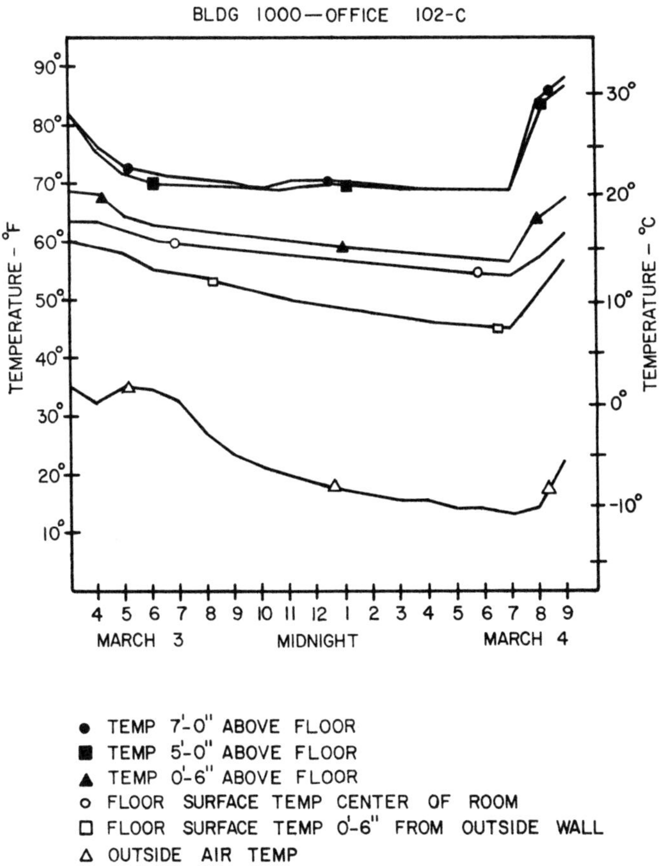

Figure 7.1. Hourly temperature distribution in an office. (From ASHRAE TRANSACTIONS, *Vol. 82, Part 2, copyright 1976. Used by permission of American Society of Heating, Refrigerating and Air-Conditioning Engineering, Inc.)*

Additional Insulation

Insulation of floors and the addition of storm windows were considered as a way to conserve energy and at the same time to improve conditions found at floor level and near outside walls. Floor insulation was assumed to be carpeting with a 1/2-inch (1.27-cm) pad.

Energy requirements after adding carpeting are identified as Case 5 in Table 7.1. Requirements after the addition of storm windows are identified in Case 6 of Table 7.1. As seen, both additions resulted in a decrease in energy used for heating and an increase in that used for cooling for a net decrease of 1.3%.

Reset Temperatures

Resetting of inside temperatures as a means of conserving energy has been widely discussed and encouraged from national down to plant levels. In Building 1000, the temperature settings adopted were consistent with those being recommended, that is, 70°F (25.6°C) in summer and 68°F (20°C) in winter.

Energy requirements subsequent to adding temperature reset are identified as Case 7 in Table 7.1. As seen, a modest decrease in energy usage was realized.

Daylight Savings Time

Daylight savings time has also been widely applied as a means of conserving energy. In Building 1000, hours of work have been adjusted in accordance with nationally published schedules--i.e., from April to October and on a year-round basis. For the purpose of this study, both schedules were considered.

Energy requirements after the addition of an April to October schedule are identified as Case 8 in Table 7.1. Requirements after the addition of a year-round DST schedule are identified as Case 9 in Table 7.1. As seen, energy required for heating increased while that required for cooling decreased. A minimal net savings resulted from the April to October schedule while a small net increase resulted from the year-round schedule.

Operating Experience

Except for the addition of carpeting and storm windows, all of the alternatives studied have been implemented. Theoretically, the energy requirements for the building on a month by month basis should be as shown in Figure 7.2. For comparison, energy usage has been monitored for the period from July 1975 through March 1976 and is also shown. The measured data were corrected on a degree-day basis to reflect the average weather used in this study.

As seen in Figure 7.2, measured energy usage was somewhat lower than predicted during the summer months. This is attributed to

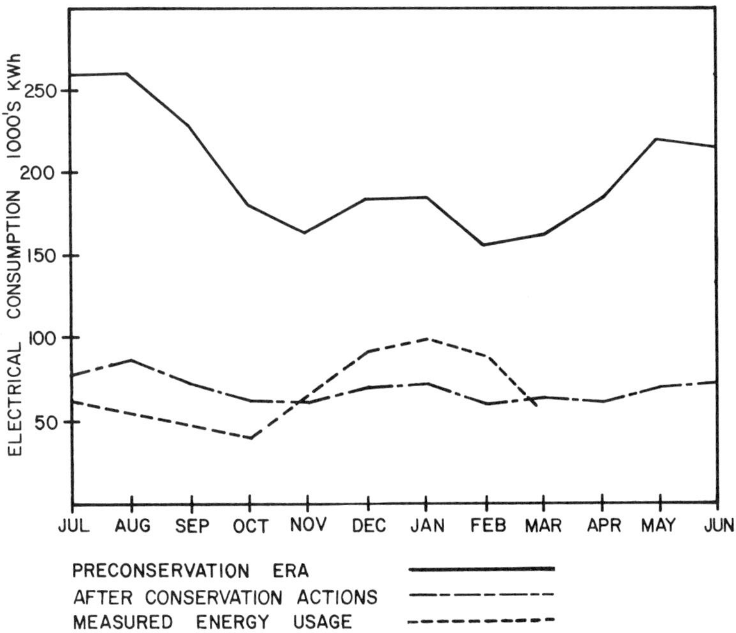

Figure 7.2. Pre- and post-conservation energy usage in building 1000. (From ASHRAE TRANSACTIONS, *Vol. 82, Part 2, copyright 1976. Used by permission of American Society of Heating, Refrigerating and Air-Conditioning Engineering, Inc.)*

lights being turned off when occupants leave at 4:30 p.m. Originally it was assumed that lights would remain on until the janitors leave at 11:00 p.m.

During December, January, and February, measured energy usage was found to be higher than expected. A random survey of offices revealed that some of the baseboard heaters were not being turned off as assumed.

Occupants of Building 1000 have experienced slight discomfort as a result of the temperature settings. No adverse effects have been noted as a result of less lighting and ventilation. With equipment shutdown overnight during the summer, offices are a bit stuffy at 8:00 a.m. but not uncomfortable. During the winter, overnight temperatures decrease as previously illustrated, but are reestablished to a comfortable level once heaters are turned on. After a

Table 7.2. Energy Consumption--Building 1000*

	Total electrical			Energy[+] costs ($/yr)	Energy saved ($/yr)
		%Δ	Σ %Δ		
Pre-energy-conservation era mode	2413.6	-	-	36,600	-
Modified lighting schedule	1860.2	-22.9	-22.9	28,200	8,400
Reduced lighting levels	1588.9	-11.2	-34.1	24,100	4,100
Reduced ventilation	1328.3	-10.8	-44.9	20,100	4,000
Equipment shutdown	842.4	-20.1	-65.0	12,800	7,300
Add first floor carpet	826.7	-0.7	-65.7	12,500	300
Add storm windows	811.2	-0.6	-66.4	12,300	200
Reset temperature, 78°/68°F	785.9	-1.0	-67.4	11,900	400
Daylight savings time--April to October	778.7	-0.3	-67.7	11,800	100
Daylight savings time--year round	783.5	+0.2	-67.5	11,900	(100)

*1000's kWh.

[+]Electricity costs (8-1-75): $15.16/1000 kWh.

weekend shutdown during extreme weather, several hours are required to reestablish comfort conditions, especially at floor level. This probably accounts for heaters being left on during the winter months.

Costs of the required modifications have been minimal. Since implementation is mostly dependent on administrative control, the cost involved has been limited to the removal of light bulbs and the installation of timers on the exhaust fans. The total cost of these items is estimated to be less than $2000.

Summary and Conclusions

A summary of all energy conservation alternatives considered along with the resulting savings or costs are shown in Table 7.2. Conclusions are made as follows:

1. Almost 45% of the energy formerly used in Building 1000, which is worth $16,500 per year, is being saved without any adverse effects and without excessive costs for modifications.
2. Another 20%, worth $7300 per year, is being saved with only slight adverse effects and without excessive modifications.
3. Additional insulation does not seem to be an attractive economic alternative but could possibly be justified on the basis of improved comfort conditions.
4. Resetting thermostats to 68°F (20°C) in winter is viewed as inappropriate when the resulting discomfort is weighed against the small amount of energy saved.
5. Daylight savings time does not seem to be worthwhile for Building 1000 in eastern Tennessee.

CASE 2: ENERGY MANAGEMENT OPPORTUNITIES IN A MANUFACTURING FIRM*

Description of the Company

Myers-Whaley Company is an old-line machinery manufacturing company engaged in the building of mining equipment for surface and underground mining. Founded over 75 years ago, it was a builder of custom, underground, coal-loading equipment. The company developed into a custom machinery builder for other types of industry as the

---

*Reprinted with permission from the Myers-Whaley Company, Knoxville, TN.

market for coal machinery declined in the late sixties and early seventies. In 1973, the company reentered the coal equipment business to market these specific products. The present product line consists of a battery-powered personnel carrier, a battery-powered rock duster car, and a diesel-powered auger, plus parts for discontinued and current products.

The company purchases electric drive motors, steering gear boxes, diesel motors, wheels, bearings, hydraulic cylinders, gears, and some formed sheet metal parts. It cuts and machines, welds, and bends steel bar stock and shapes for the chassis and gear housings. It assembles drive transmissions for the auger miner from purchased gears. It assembles both purchased and produced parts and subassemblies, and paints and tests the completed machine. It also produces experimental models of machinery developed by its R&D engineering staff.

The company operates in three buildings with a total area of approximately 31,000 ft^2 in an industrial section of Knoxville, Tenn.; see Figure 7.3. The oldest building is a 23,000-ft^2 wooden beam structure with three floors in the center of the roof apex. The second floor is utilized for an office and the third for dead storage. The smaller structures are of similar construction, but are single floor, high ceiling manufacturing areas. Building 1 houses purchased parts storage in the front. Assembly operations are performed for all products in this building. Machine shop equipment is located in the areas designated A, B, C, D. Building 3 is used for steel cut-up. Building 2 is used for frame welding and painting. A new building is being planned for the painting operations to comply with OSHA and pollution standards.

Presently there are 40 employees, but at peak the total may increase to about 70. The shop employees are on a four-day, 10-hours per day, one-shift schedule of 40 hours per week. Office employees work eight hours per day, five days per week.

The company has two executives; one handles sales and production and the other attends to financial matters. A receptionist-

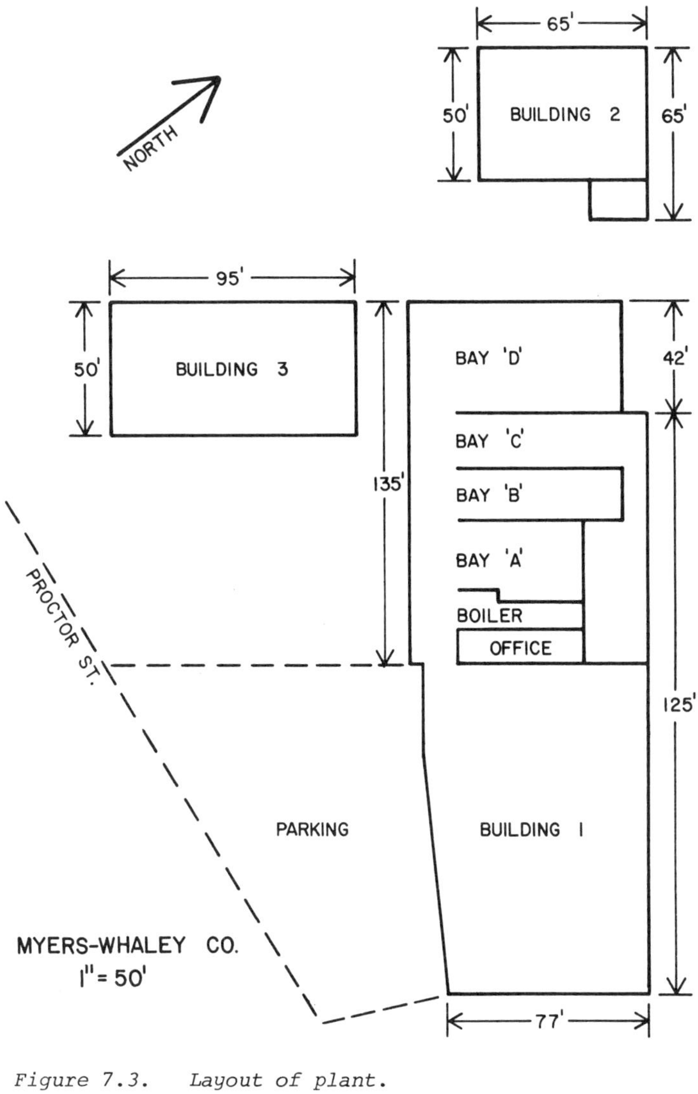

Figure 7.3. Layout of plant.

secretary, an accounts-payable clerk, and an engineer-draftsman comprise the main office staff. See Figure 7.4 for an organization diagram.

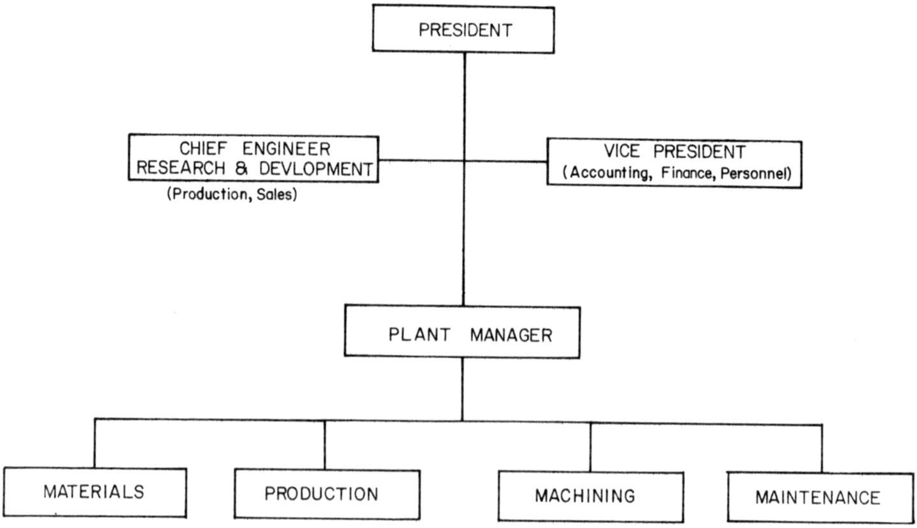

Figure 7.4. Organization of company.

Objectives of the Study

The objectives of the study are as follows:

1. To identify direct energy uses and flows in each operation of the manufacturing process of the firm
2. To analyze the uses and flows of energy to determine the effectiveness of energy management
3. To determine and evaluate energy conservation opportunities
4. To develop an energy management plan for implementation of worthwhile conservation opportunities and to establish a continuing conservation program
5. To identify and suggest implications of the energy shortage on long-range strategy

The study has been limited to direct energy purchases (gas and electric) in order to fit the scope of the study to time and personnel constraints. It is suggested that management analyze the energy content of raw materials and purchased parts with the objective of evaluating possible material substitutions having lower energy content.

Organization of the Study

The study was organized into six areas:

1. Management-employee awareness
2. Lighting use and effectiveness
3. Space heating (combustion and distribution)
4. Compressed air (production and distribution)
5. Electric motor and equipment survey
6. Overall energy pattern consumption

With regard to *management-employee awareness*, management interest and intent were established in a preliminary interview between the chief operating officer, a team member, and the area representative of the Industrial Services Division of the University of Tennessee.

Making a plan work in any organization depends on two factors: (1) complete and obvious support of top management, and (2) acceptance and support of employees. The relatively simple three-tier organization structure makes the management line of communications short and easily controlled. The demonstrated cooperative attitude of the employees makes the communication problem minimal. The fact that management and employees worked out a mutually beneficial energy saving work schedule of 10 hours per day, four days per week, at the beginning of the energy crunch demonstrates this attitude. Both management and employees are satisfied with the schedule.

To implement an ongoing energy management program, we suggest the following. The first decision to be made is where to place the responsibility for the energy conservation effort. It should be someone with the technical competence to evaluate opportunities and also someone who would stand to benefit from any savings which would accrue from conservation. In this particular situation, the plant manager is the logical person to be given energy management responsibilities.

The second step is to inform the employees of the energy management program. Communication in this company is usually handled by either posting an announcement on the bulletin boards or by calling all the workers together and making the announcement.

In this case, since the program will require the understanding and support of the employees, it would probably be best to have a plant meeting and have the executives of the company announce the program, and also to post the announcement on bulletin boards.

The third step would be for the plant manager to enlist the help of two or three operating people to form the energy conservation committee.

Lighting use and effectiveness surveys measure the available and utilized lighting against norms established for each type of area and kind of work being performed in the area. In addition, the use of lighting levels is considered in terms of supplemented daylight contribution to level of need.

A *space heating survey* begins with the input fuel supply, measures the efficiency of the combustion process and heat distribution system, and then identifies and measures losses. Energy inputs are balanced against outputs and losses.

Compressed air production and distribution surveys look at power input source, compressor efficiency, distribution piping, and point of use to determine useful work contributions and losses.

Electric motor and equipment surveys identify each motor or other electric energy-consuming device, determine consumption per unit of time, and establish a usage factor based on production level and work schedule. Electrical energy purchased is balanced against computed consumption.

Overall energy consumption pattern analysis involves the determination of seasonal or time cycle consumption patterns, consumption trends, total energy costs, and unit energy costs, and the analysis of data to establish reasons for variations. Since energy costs have peak demand as a major cost determinant, the examination of patterns and trends may lead to cost reduction and long-range energy capacity conservation.

In order to maximize the effectiveness of the team members, specific areas of audit responsibility were assigned to the individuals on the team as follows: R. D. Chambers--management and employee awareness, and lighting energy use effectiveness; W. Pflager--space heating survey, and compressed air production and distribution; W. Redford--electric motor and equipment survey, and direct energy purchases and trends.

The Energy Audit

The audit proceeded by identifying the process flow and energy consumption in each segment of the manufacturing process, overall consumption and cost by type, combined energy consumption and cost, and consumption per unit of production.

Product Flow Chart and Energy Consumption

Figure 7.5 is a flow chart which identifies the major processes involved in the manufacture of the three principal products:

1. Battery-operated mine personnel car
2. Battery-operated mine rock duster
3. Auger miner (surface mining)

The operations are not sufficiently separated in manufacture to determine the energy consumption for each product, but the installed capacity for each process has been determined from an energy audit conducted in February 1976. Included also is an estimated use factor for each process and a projected monthly energy consumption per process based on the installed capacity and use factor for one shift per day of 10 hours, four days per week production schedule.

Figure 7.6 is a plot of monthly electrical energy consumption and natural gas consumption for the period January 1973 through December 1975. The gas consumption billed in therms (100,000 Btu) is converted to kWh equivalent. The obvious seasonal consumption for gas indicates its use to be only for space heating.

Figure 7.7 is a plot of monthly electrical energy cost and gas cost for the same time period. The demand charge for gas which applies to the summer months May through October (the nonheating season) has been assessed against the months of use, using one-sixth of the total demand charge as an additional cost for the

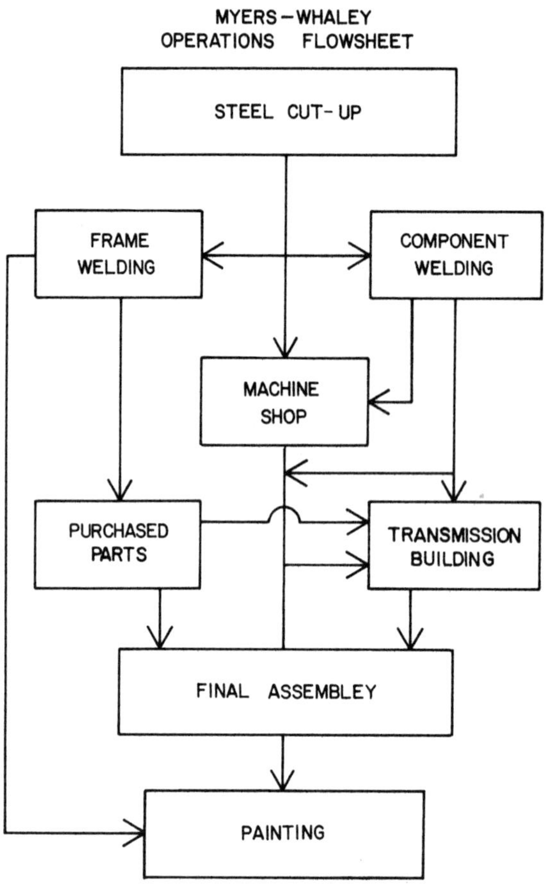

Figure 7.5. Operations flow sheet.

months in which gas is used. The demand charge for gas is based on 5% of the highest monthly consumption during the preceding November-April period and is levied at the applicable rate for 80% of that 5% demand factor.

The extreme variability of the electrical energy cost reflects a widely fluctuating demand level upon which a substantial portion of the charges is based. See the Appendix for an explanation of the basis of demand charge and Figure 7.11 for a typical demand chart.

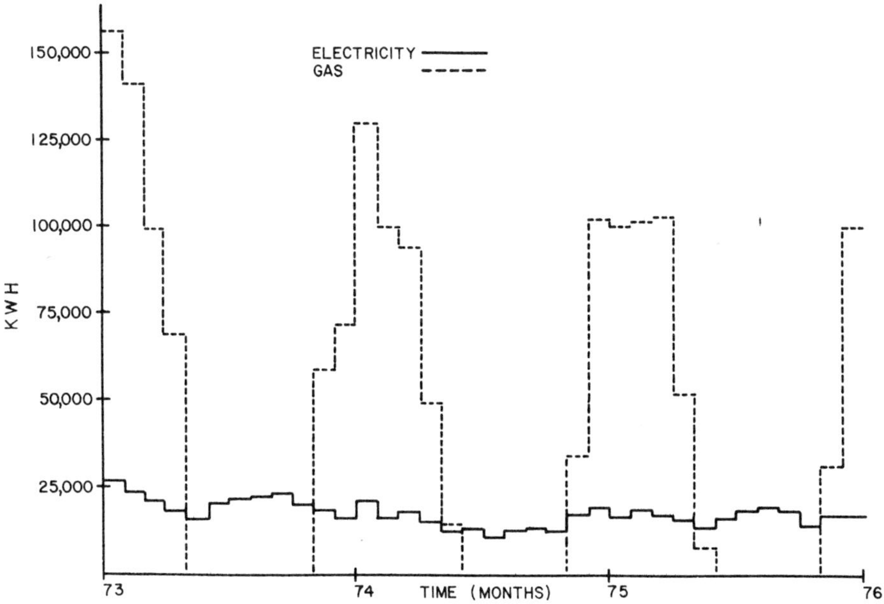

Figure 7.6. Monthly energy consumptions (kWh or kWh equivalent).

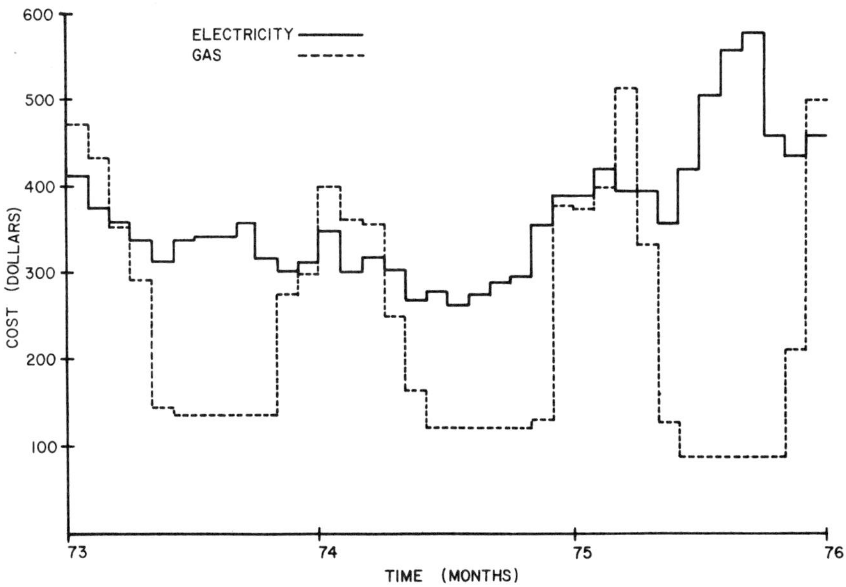

Figure 7.7. Monthly energy costs.

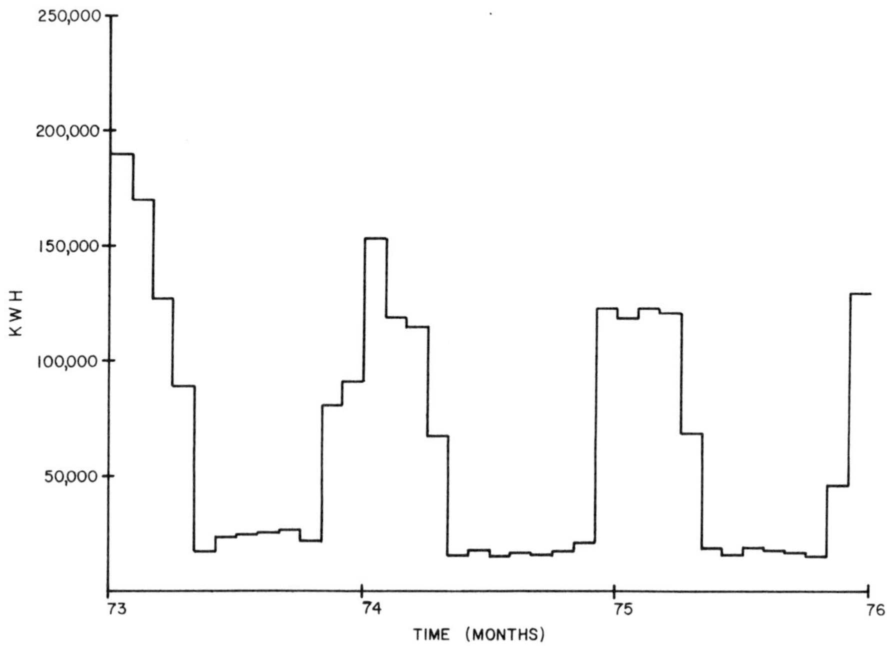

Figure 7.8. Total monthly energy consumption.

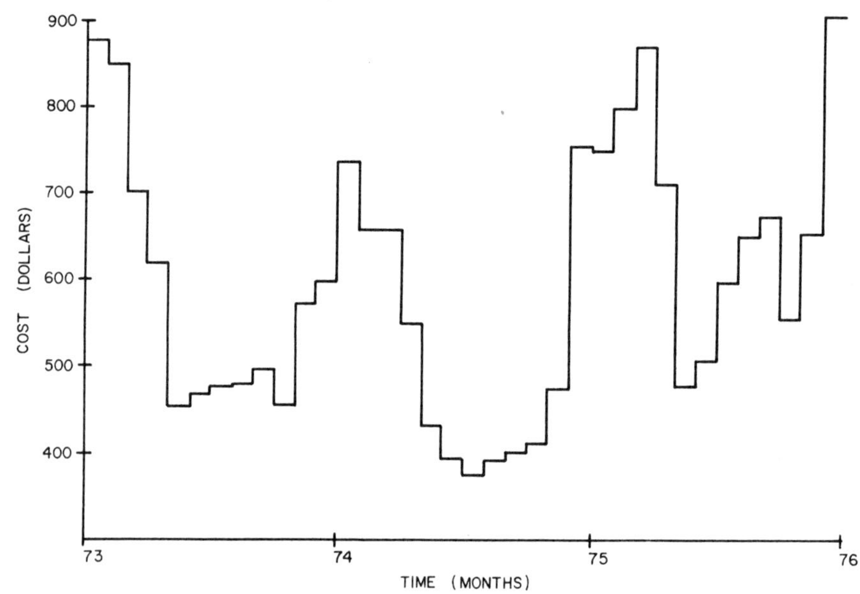

Figure 7.9. Total monthly energy cost.

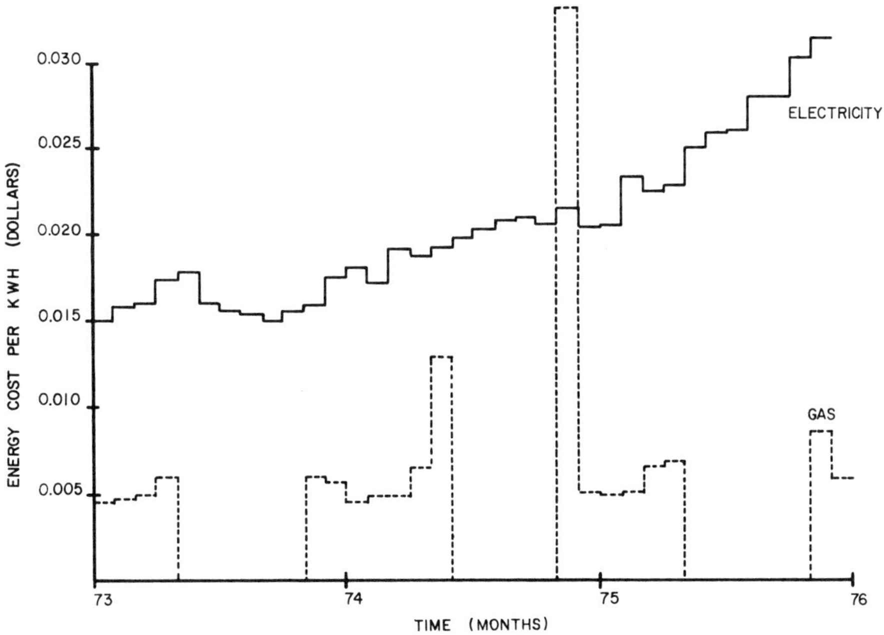

Figure 7.10. Energy cost per kWh equivalent.

Figure 7.8 combines the total gas and electrical energy consumption. The significance of the effect of gas consumption as the overriding energy source in terms of the total energy is apparent.

Figure 7.9 is a plot of the total monthly combined costs of gas and electrical energy. The combined variability on a seasonal basis is apparent.

Figure 7.10 is a plot of unit cost. The rising trend in cost per unit of energy is apparent. The increase in 1975 over 1973 cost for electrical energy was 52.5% and for gas was 24.4%. The need to conserve energy from a cost standpoint is readily obvious.

On the basis of 1975 production data regarding steel consumed in each product, as well as observation of the process, it was concluded that electrical energy consumption per unit is as follows:

Battery-operated mine personnel car, 750 kWh
Battery-operated mine rock duster, 800 kWh
Auger miner, 7500 kWh

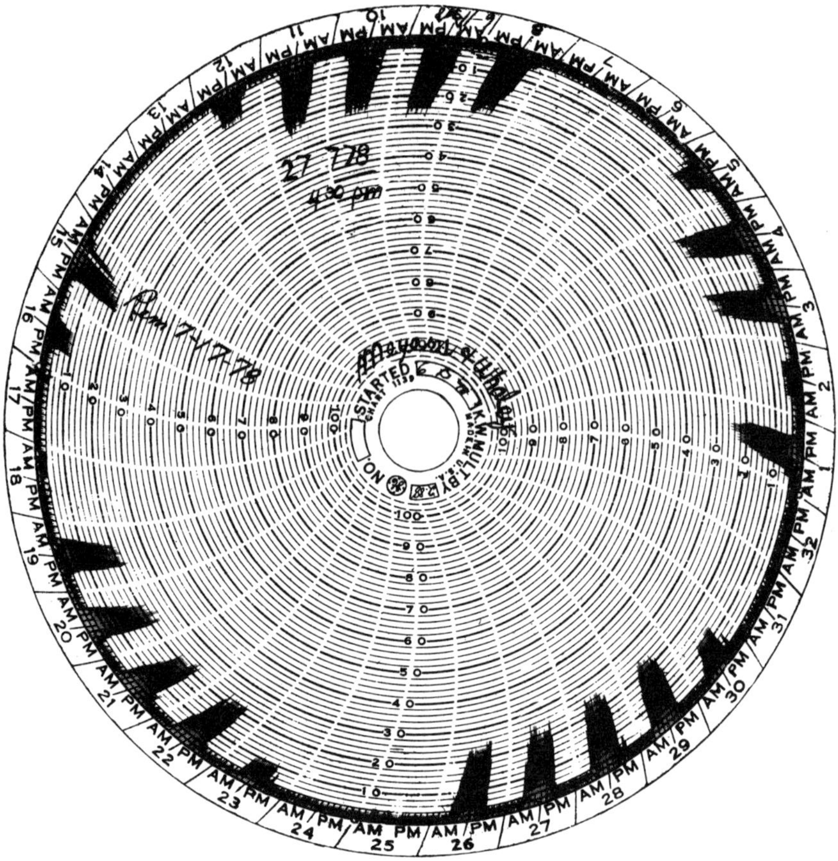

Figure 7.11. Electrical demand record. (Reprinted with permission from the Myers-Whaley Company, Knoxville, TN.)

If gas consumption is allocated in approximately the same proportion (and this would seem to be an equitable manner), the gas consumed per unit is as follows:

Battery-operated mine personnel car, 1900 kWh equivalent
Battery-Operated mine rock duster, 2000 kWh equivalent
Auger miner, 19,000 kWh equivalent

Energy Conservation Opportunities (ECOs)

In evaluating ECOs it is advantageous to divide them into two major areas: (1) ECOs by deprivation--that is, ECOs that can be implemented with some detriment to the company, in personnel comfort, competitive advantage, etc.; and (2) ECOs by improved utilization efficiency. The latter are ECOs which can be accomplished at little or no personnel discomfort and which should improve the company's competitive advantage.

The most obvious areas in which to conserve by deprivation are heating and lighting. But these areas were not feasible candidates because of possible violations of OSHA regulations and also because of increased assembly time and error rate due to personnel discomfort and low lighting levels. Effective utilization of heat and light will be discussed later.

A minor deprivation ECO that was considered was the disconnection of water coolers and refrigerators. This was not judged feasible because the total annual cost of electricity for these appliances was under $50 and it was felt that potential savings in this area would be outweighed by the inconvenience and alienation of the employees.

The ECOs by efficient utilization involve a number of areas. These can be grouped into two categories. The first category contains measures that can be implemented by procedural or scheduling means, i.e., ECOs that require little or no capital investment. The second category involves ECOs that would require a capital expenditure.

Procedural and Scheduling Changes

Reduction of demand charge for electrical energy The area which presents the most opportunity for savings is in efforts to reduce the demand charge. Based on an analysis of demand charts for each month of 1973, 1974, and 1975 (a sample of which is shown in Figure 7.11), we are confident that the cost of electrical energy can be reduced by the equivalent of the charges for 100 kW of demand. This will result in an annual savings of \$2,400 determined as follows:

100 kW demand saved × \$2.01/kW demand rate = \$201/month

Annual saving = 12 × \$2.01 = \$2412

The method of achieving this level is by scheduling welding operation so that a limited number of welders are in use at the same time since welders constitute the largest single variable demand source. One welder coming on line will push demand up by 5 to 40 kW.

Scheduling building usage A first impression of the buildings and grounds of Myers-Whaley is that there are many large rooms which are utilized on an occasional basis. This would suggest that through proper scheduling energy could be saved by working in certain rooms on certain days. Closing individual rooms off is not practical due to the building layout. An alternative method, though, since the company has three buildings, would be to schedule the two smaller buildings for three days a week, i.e., a 25% reduction in building utilization. If this were done, the savings would be as follows:

Heating: Building 2 contains 20% of the total floor area heated by the boiler:

$$\begin{Bmatrix}\text{Annual boiler}\\ \text{fuel bill}\end{Bmatrix} = \begin{Bmatrix}\text{total annual}\\ \text{gas bill}\end{Bmatrix} - \begin{Bmatrix}\text{bill for gas used}\\ \text{in direct heating}\end{Bmatrix}$$

$$= \$(2975 - 338) = \$2637$$

Annual heating bill for building 2 = 20% of \$2637

= \$537

Reduction of this bill by 25% gives a saving of 25% of \$537 or \$134 per heating season.

Building 3 is heated directly by gas heaters using an estimated average of 75,000 Btu/hr. The reduction of 10 hours per week (25%) in building use will result in a savings of:

75,000 Btu/hr × 10 hr/wk × 24 wk/heating season

= 18,000,000 Btu/heating season

At the current rate of $1.50 per million Btu for gas, the saving amounts to $27 per heating season.

Lighting: Building 2 has 4.44 kW of installed lighting. The running time will be reduced by 10 hours per week, reducing the annual electric bill by the following amount:

$$4.44 \text{ kW} \times \frac{10 \text{ hr}}{\text{wk}} \times \frac{50 \text{ wk}}{\text{yr}} \times \frac{\$.03}{\text{kWh}} = \$67/\text{yr}$$

Building 3 has 3.60 kW of installed lighting capacity. The reduction in annual bill will be

$$3.60 \text{ kW} \times \frac{10 \text{ hr}}{\text{wk}} \times \frac{50 \text{ wk}}{\text{yr}} \times \frac{\$.03}{\text{kWh}} = \$54/\text{yr}$$

Total saving from this ECO:

Heating ($161) + lighting ($121) = $282/yr

Lower air compressor pressure Lowering the pressure of compressed air is another way to save energy. The foremen were asked if it would be possible to reduce the air pressure and their judgment was that it would be acceptable. Lowering the pressure from 110 to 95 psi would mean a 5.4% saving, based on handbook data [2] and calculated as follows:

$$\text{Energy saving} = 10 \text{ hp} \times 0.746 \frac{\text{kW}}{\text{hp}} \times 5 \frac{\text{hr}}{\text{day}} \times \frac{4 \text{ days}}{\text{wk}} \times \frac{50 \text{ wk}}{\text{yr}}$$

$$= 0.054 = \frac{403 \text{ kWh}}{\text{yr}}$$

At $0.03/kWh this saving is currently worth $12 per year.

Analyze boiler flue gas The flue gas in the boiler was analyzed to check for excess oxygen and to measure stack gas temperature. Two percent oxygen and 550°F stack gas were found. The conclusion is that the boiler is operating efficiently.

ECOs Requiring Investment

Repair leaks in compressed air lines An inventory of the compressed air system showed 33 outlets, nine of which were leaking. Estimating that these leaks are the equivalent of a 1/32-in.-diameter hole and assuming the system is pressurized 40 hours per week and using handbook data [3] rated for 100 psi (which would give a conservative estimate for the actual system), the savings per year at \$0.03/kWh would be \$183. Cost of material to replace hoses at all outlets would be about \$140 (33 outlets × 25 ft/outlet × \$0.17/ft).

Reduce heating In this area the most necessary action needed is the institution of control. Every time the plant was visited there were some examples of waste which could be avoided. For example, the gas heaters in building 3 were usually running in the afternoon if it had been cold in the morning. At times the heaters would be running and doors and windows would be open. On one occasion the heater fans in building 2 were operating when there had been no heat necessary for several hours.

The main thermostat has no temperature setting on it (only numbers 1-5); it is sometimes set on high (4) in warm weather and runs unnecessarily. A solution to this problem would be to determine what temperatures corresponded to the marks on the thermostat, set the thermostat at a lower temperature, say 68°F, and put a cover over the thermostat to prevent unauthorized setting changes.

For building 2, where the heater fans are controlled by manual switches, the best alternative would be to wire the fans so that they run only when the main system is operating, which is the arrangement presently used in the main building. The savings from such a move would be difficult to estimate, but with the low investment required the cost is obviously worthwhile.

In building 3, which uses direct fired gas heaters with manual controls, a thermostat should be installed for better control. The savings are calculated as follows (assume the thermostats would reduce the use of the heaters by an average of 2 hours per day):

$$\text{Gas used at a rate of } 8(\text{heaters}) \times 30{,}000 \text{ Btu/hr} = 240{,}000 \text{ Btu/hr}$$

The reduction in use time over a complete heating season is

$$\frac{2 \text{ hr}}{\text{day}} \times \frac{4 \text{ days}}{\text{wk}} \times \frac{24 \text{ wk}}{\text{heating season}} = \frac{192 \text{ hr}}{\text{heating season}}$$

The savings resulting from reduced use of the heaters is, therefore,

$$\frac{240{,}000 \text{ Btu}}{\text{hr}} \times \frac{192 \text{ hr}}{\text{heating season}} = \frac{46 \text{ million Btu}}{\text{heating season}}$$

At the present rate of $1.50 per million Btu for natural gas, the saving is worth some $69 per heating season.

The cost of a thermostat would be about $25 installed, so the investment would pay for itself in the first few months of operation.

Control lighting levels to avoid excess lighting Due to the many windows in the building there are times when natural light is adequate for several areas. For the general assembly work that is done in most areas of the shop a lighting level of 30-50 footcandles is adequate. Lighting measurements were taken in the shops on a sunny afternoon and the lighting levels were from 22 to 200 footcandles. There are several ways to control the operation of the lights, such as using timers or photocells, or making one person responsible for light levels. Timers are cheap and simple but do not have the flexibility to adjust to varying seasonal available light, overcast days, etc. It would not be practical to put a person in charge of lighting levels full time, so it would have to be made a secondary responsibility of a present employee. This would probably be feasible for a short time but over long periods of time secondary responsibilities tend to be forgotten. For cost and dependability a photocell system would seem to be best. A photocell to control several sets of lamps would cost about $35 installed (this includes $15 for parts and $20 for labor). For estimating purposes, say, the natural light would be sufficient for five hours a day for six months a year for a total of 520 lighting hours saved. The large mercury vapor lamps in the main building and the two auxiliary buildings are candidates for this application. Using three photocells to control the mercury vapor lamps in the three buildings would cost a total of $105, with a savings of about $187 per year (520 hr × 12 kW × $0.03).

Insulate buildings At present the buildings have little insulation. According to a commercial insulating company (Godfrey

Insulating Company), it would take about \$16,250 for first class insulation, which should cut heating expenditures between 30% to 40%. This is a savings of \$900-\$1200 per year. At the present cost of gas it would take over 40 years to recover the investment, using the net present value method of evaluation.

The calculations are as follows:

\$1200 (savings per year)   9779 (NPV, 40 periods at 10%)
  = \$11,735

Unless the price of natural gas increases or the price of insulation decreases, this ECO will not be justified.

Return condensate to boiler system Condensate steam is presently being expelled at one end of the building at the rate of 5 gallons per hour at almost 212°F. This translates to an annual cost as follows:

$$\text{Energy in condensate} = \frac{5\ \text{gal}}{\text{hr}} \times \frac{8\ \text{lb}}{\text{gal}} \times 160°\text{F} \times \frac{1\ \text{Btu}}{\text{lb-}°\text{F}}$$

$$= 6400\ \frac{\text{Btu}}{\text{hr}}$$

In a full heating season the energy loss is

$$\frac{6400\ \text{Btu}}{\text{hr}} \times \frac{6\ \text{hr}}{\text{day}} \times \frac{4\ \text{days}}{\text{wk}} \times \frac{24\ \text{wk}}{\text{heating season}} = \frac{3.7\ \text{million Btu}}{\text{heating season}}$$

which at current gas prices is worth

$$\frac{3.7\ \text{million Btu}}{\text{heating season}} \times \frac{\$1.5}{\text{million Btu}} = \$5.50/\text{heating season}$$

The cost of the water itself is about \$8.50. Total cost per year is \$14.

A package to collect and pump the condensate back into the system costs approximately \$700 installed. This ECO is not practical at present due to poor cost/benefit ratio.

Close and seal external openings There are numerous small holes in the walls and doors, especially in the smaller buildings, which would take little to repair and would save heat. One major example of this is a stack in building 3 which is not used at present but is still open. This allows a considerable amount of heat to escape. There are also several windows that do not close properly. These should also be repaired. This last ECO would cut heat loss from the buildings, but it is difficult to estimate the amount.

Summary of ECO Recommendations

This section summarizes the actions recommended in the preceding analysis.

ECO	Recommendations	Annual savings
Reduce demand charge for electrical energy	Implement immediately	$2412
Lower air compressor pressure	Implement immediately	224
Schedule building usage	Implement immediately	282
Repair leaks in compressed air lines	Implement immediately	(1st year) 43 (2nd year) 183
Control lighting levels	Implement immediately	(1st year) 82 (2nd year) 187
Reduce heating	Implement immediately	(1st year) 44 (2nd year) 69
Close external openings	Implement immediately	Not estimated
Insulate building	Not recommended at present; should be monitored in relation to fuel costs	
Return condensate to system	Not recommended	
Analyze flue gas to assure efficient combustion	Already done	

From this summary it can be seen that the company can save about $3092 the first year and $3357 in the following years by adopting the recommended ECOs.

Strategic Implications of the Energy Shortage

The following areas of conservation will require intensive study, development of alternative proposals, and evaluation in the perspective of the long-range goals of the company.

Natural gas substitute The option to choose natural gas as a source of energy for space heating is not likely to continue to be available at a competitive price. It is our highest quality fossil fuel and the most useful source of hydrocarbons for specialized use in metal finishing and glass production. The continued misuse of gas for heating is likely to be discouraged by proscriptive regulation or price deregulation to allow the market to determine the price by comparison with other fuels.

Plans should be made to convert to another fuel at an opportune time, i.e., when its use is restricted, its price prohibitive, or the present boiler needs replacement. The use of electricity is not now economical at its present incremental cost, which is approximately three times that of gas. However, the differential is likely to decrease. Also, consideration could be given to heating work personnel directly rather than the entire space in the building.

Value analysis for make-buy decisions The cost of a component in terms of total cost and energy content should be determined for each part produced and for the same part if purchased from an efficient producer. It would appear that a number of components that are now intricate stress-relieved weldments might be forgings, or machined, heat-treated castings. Components with multiple holes might be produced more economically on a multiple head drill press than on a single head press. The volume required may not yet make the option of buy rather than make realistic at this time, but as production of standardized parts increases, the need to regularly and critically evaluate the make or buy decision for each item is apparent.

Streamline production flow The growth and development of the product line around a custom fabrication shop is evident. While some assembly-line techniques have been adopted, more are in order. Space utilization should be more efficient. Though it is not within the scope of this investigation to develop a production layout, we believe that by moving out unused equipment, rearranging production flow, and placing a paint booth in an existing building, the need for a new building for a paint shop, with its attendant energy costs, can be avoided. A thorough study of flows and material handling techniques by an industrial engineer is suggested.

Product line Energy conservation is not an end in itself. The desired end result is, for society, more efficient use of natural resources, and, for the company, improved competitive position. However, the most efficient company will not survive if there is no longer a need for the company's products. Since this company has a specialized market and a somewhat limited product line, it should make every effort to predict future demands for coal, the industry to which the company is tied at present. It should also attempt to predict whether the present ratio of strip versus surface mined coal will remain stable or not, and if not, in which direction it is likely to change. With this information the company could decide whether to diversify and also know where to place the product development emphasis.

Although this study did not probe the financial aspect of the company's operation, the financial situation should be analyzed to be sure that the capital that will be required for future opportunities and problems will be available.

Appendix. Demand and Energy Charges in the Monthly Electric Bill

In order to supply electric power to a customer, a utility company incurs two different types of expense. The first is the cost of facilities and equipment required to deliver the power; the second is the cost of the energy used. The cost of energy used is paid directly by the customer as part of the energy charge. Part of the remaining cost is fixed (for example, the costs of meter reading, accounting, service, and information) and is made a part of the energy charge. The remainder is the cost of installing and maintaining the equipment (transformers, distribution lines, and circuit breakers) necessary to bring power to the customer. This cost will depend on what size equipment is required to supply the particular load and, therefore, it depends on the *maximum* rate at which energy is drawn from the supply system. For this reason, the utility company applies a demand charge, determined from the *maximum* demand (or power drain) for each billing month. The demand is determined by recording the kWh (energy) used in each 30-minute (sometimes 15-minute or 60-minute) period. The *largest of these readings in the month* determines the maximum demand and hence the demand charge for the month.

The demand charge for Myers-Whaley was $2.01/kW (for highest category) in February 1976. This amount can be saved by reducing the maximum demand in any month. It is important, therefore, that someone be able, on a monthly basis, to examine the utility's demand charts to look for opportunities in demand reduction.

CASE 3: INDUSTRIAL ENERGY CONSERVATION STUDY IN A GLASS-DECORATING PLANT

This report is one of a series resulting from a program sponsored by the U.S. Department of Commerce, through the University City Science Center, Philadelphia, PA, to evaluate the concept of an Energy Analysis and Diagnostic Center (EADC) as a focus for energy conservation resources for small and medium sized companies. This is one of some 25 reports prepared as the result of plant visits by project personnel at the University of Tennessee. Some of the details of the company recorded at the beginning of the report were required in the concept evaluation process and are not related directly to the energy management aspects of the report itself. They are, however, of general interest in establishing the nature of the company involved.

The objective of the reported survey is to identify potential energy conservation opportunities. The provision of engineering design services is outside the scope of this particular project.

As a result, the cost of realizing the identified potential, in those cases where capital investment is required, is not determined. The investment information is, however, given in terms of the investment that can be made when the payback period is known. In this way the company can establish from its own payback criteria an appropriate level of capital investment and can investigate the engineering feasibility, either with in-house staff or outside consultants, of realizing the potential savings under that investment constraint. The limitations of the payback method are neglected in this report and the payback formula used on page 158 is a well-established [4] modification of the one given in Chapter 4, taking into account the time value of money.

Description of the Company

Type of Industry

The company manufactures tempered glass plate from glass stock, principally of 3/16-inch thickness. The manufactured pieces are used as decorative glass panels in residential appliances and as fire screens for fireplaces. The basic raw material is float glass plate stock and the decorative process involves silk screening.

Number of Employees

The company employs 68 hourly wage employees.

Operating Characteristics

The normal operating schedule is two eight-hour shifts, five days per week.

Type and Yearly Production of Goods

Because of the wide variety of decorative glass plates produced, the production output of the company is measured in terms of gross sale value. The amount of scrap material generated varies from 15 to 30%, depending on the complexity of the piece. Total gross sales for 1976 were $1,225,352 distributed on a monthly basis as shown in Table 7.3.

Table 7.3. Monthly Gross Sales for 1976

Month	Sales
January	$ 96,711
February	99,570
March	121,920
April	80,547
May	104,048
June	110,676
July	78,551
August	99,005
September	116,215
October	92,514
November	112,153
December	113,442

General Description of Company Services

The company's general market is the major appliance manufacturers in the United States. It sells tempered and silk screened glass panels for use in ranges, clothes washers, clothes dryers, etc. In addition, the company does some specialty glass tempering for a varied clientele on a bid basis.

The tempering/silk screening operation is a highly sophisticated procedure and, therefore, only two companies in the United States are in direct competition for the bulk of this business. The specialty tempering portion of the business (a small fraction) is a highly competitive field where many companies have the manufacturing capability to bid on orders.

Energy Utilization in the Company

Role of Energy in the Industry

The SIC code for the company is 3231, glass products made of purchased glass. On a national basis, the industry uses natural gas and electricity.

Table 7.4. Energy Consumption per Dollar Value Shipped for SIC 3231

	National	Company	National	Company
Natural gas	5.69 ft^3/$	8.50 ft^3/$	5,690 Btu/$	8,500 Btu/$
Electricity	0.48 kWh/$	2.04 kWh/$	1,638 Btu/$	6,972 Btu/$
Coal	0.019 lb/$	--	266 Btu/$	--
Total			7,594 Btu/$	15,472 Btu/$

The energy consumption for SIC 3231 per dollar value of shipment is shown in Table 7.4. The national data were taken from the 1963 Census of Manufacturers General Summary. This company uses only natural gas and electricity with no alternate fuel for the natural gas system currently available.

Prime Energy Consumption and Costs

Figure 6.6 shows the monthly electricity consumption and cost for 1976. Figure 7.12 shows the same data for natural gas. Figure 6.7 shows the ratio of total energy consumption (electricity plus natural gas) per dollar value of gross sales.

Prime Energy-Consuming Systems in the Company

The decorating and tempering of glass panels require a large amount of manual handling. The cutting, grinding, drilling, and silk screening processes are all done by hand in this plant. The tempering of the glass in the annealing furnace is by far the single most significant energy-consuming process. The annealing furnace consumes natural gas continuously at a rate between 1.0 and 2.4 million Btu/hr and runs continually throughout the week (three-shift operation). It is currently the production bottleneck. It is shut down for maintenance during the weekends (24 hours).

The second largest energy user is an electrically heated Lehr furnace used in the special treatment (1000°F) of precious metals used as decoration on some glass panels. The rating of this device is 200 kWh (680,000 Btu/hr) but it appears to be well insulated.

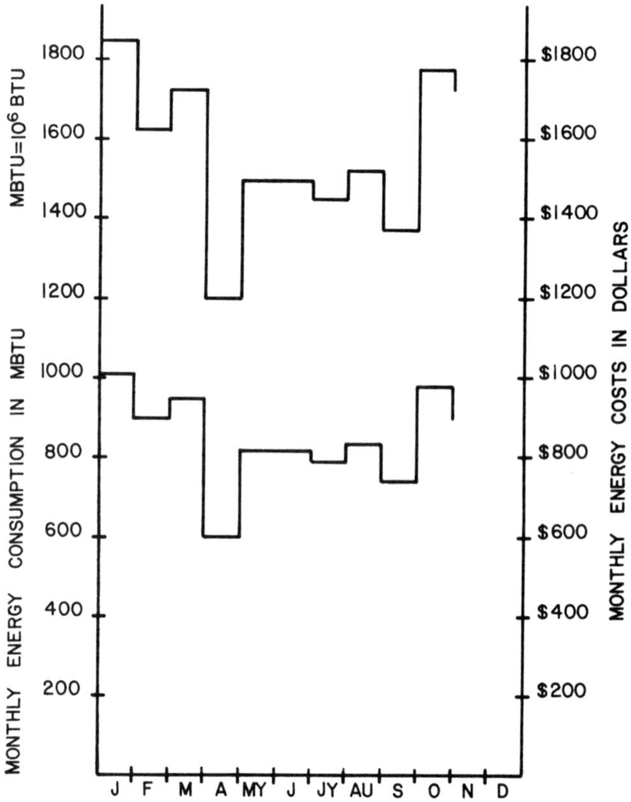

Figure 7.12. Natural gas energy consumption, year of 1976.

The major heat loss is probably through the use of the purging system. Since this equipment was not in use on the date of the visit it was not possible to make detailed observations of the operation.

Of secondary but measurable importance was the drying oven, rated at 90 kWh and used continually on a batch loading basis throughout the two shifts of each working day, and the shop floor lighting, consisting of an installed capacity of some 50 kWh of mainly mercury lamps.

Space heating energy needs are not great, being furnished by two units--one for the front office complex and the other for the silk screening shop (where filtered clean air is essential).

Motivation for Energy Conservation

The company had three primary reasons for being interested in the services of the University of Tennessee EADC.

1. Economics: The company wished to lower its energy cost since this was becoming a significant portion of the total product cost.
2. Conservation: The company was on a noninterruptable natural gas contract. However, the severe winter brought out fears in the management personnel of natural gas curtailment due to shortages. Therefore, they wished to conserve as much natural gas as possible.
3. Training: The company wished to set up an energy conservation program of its own and wished to use the EADC study results as a guide.

Current Company Energy Conservation Program

The plant is less than three years old and consequently very little has been done in the way of capital changes to achieve energy conservation. Some of the larger motors have been modified to run on 440 instead of 110 volts. The effects of this are probably lost in the normal variations in monthly bills.

A more important contribution has been in procedural changes and routine maintenance. Specifically, the repair of water leaks and the shutting down of unused equipment during work breaks has been pursued by the manager of manufacturing.

Apart from these considerations the plant is being operated as built. This in itself has a positive aspect since some of the original design features are energy conserving.

Specifically of note is the exhaust system in the silk screening shop. Pine oil fumes are exhausted from the work area by duct work connected to a fan on the roof. Make-up air is provided by another duct system that brings outside air down to the work area, where it passes over the screens and back into the exhaust hood system. This helps minimize the exhausting of conditioned air, yet provides the necessary ventilation.

Also noted was the existence of submetering on the fuel and air supplies to the annealing furnace. Floscope flow meters permitted the reading of both quantities continuously. A check of the fuel-air ratio using these meters indicated that no excess air was being used. The ratio could be adjusted, and was indeed varied during start-up, by a simple control. In this particular system the air and fuel were mechanically mixed prior to delivery to the burners.

A study of the monthly utility bill data indicates electrical demand of the order of 550 kW. Assuming the energy (approximately 200,000 kWh per month) was used entirely and uniformly during the five 16-hour workdays per week, the demand would be 575 kW. From this it is concluded that the demand curve must be fairly flat over the operating part of the day and that little opportunity for demand reduction will be found. (The demand charge is 25% of an average monthly bill.) Further analysis of this aspect is not easy for this plant since the utility company reads demand from a maximum demand indicating meter and no time chart or other record of demand exists.

Engineering Recommendations for Energy Conservation

Five promising ECOs were identified and are documented individually in the following. In analyzing the economic aspects of each ECO, a relationship between payback period and capital cost was established by using the following formula.

$$n = 12\,\frac{\log[(S/rC)/(S/rC - 1)]}{\log[1 + r]}$$

where

C = capital cost

S = annual savings

r = interest rate

n = number of months to achieve payback

Because of the time limitations of the study, an analysis of the capital cost required to implement each ECO was not made. Instead the foregoing formula was solved for C to determine the level of capital investment possible to obtain a prescribed payback period n at interest rate r. The expression for C is

$$C = \frac{S}{r}\left[1 - \frac{1}{(1 + r)^{n/12}}\right]$$

For each ECO, an estimate of S was made. Payback periods ranging from 6 to 36 months in increments of 6 months were assumed and interest rates of 8 and 10% were used.

The documentation of the ECOs follows a standardized format, consisting of:

1. Identification of ECO
2. Description of current practice
3. Observations
4. Recommended action
5. Anticipated savings
6. Economic analysis

Each ECO is written up to stand alone, so certain material will be repeated in the text to insure completeness of each separate ECO.

Waste Heat Recovery Potential

Four of the ECOs involve waste heat recovery from the annealing furnace flue gas. The total annual savings possible for these four ECOs is $4062. The total capital investment possible to obtain payback in a specified number of months at interest rates of 8 and 10% is shown in Table 7.5.

Table 7.5. Payback Period versus Capital Cost for Heat Recovery ECOs

Payback period (months)	Capital investment	
	8% interest	10% interest
6	$ 1,939	$ 1,939
12	3,785	3,693
18	5,539	5,447
24	7,201	7,016
30	8,862	8,586
36	10,431	10,155

ECO 1: Heating and Ventilating in the Grinding/Drilling Area

Current Practice The grinding and drilling area is ventilated by means of louvered windows in the wall separating it from the annealing furnace area, which is, in turn, equipped with three roof-mounted exhaust fans. It is heated, when necessary, by means of gas-fired, ceiling-mounted, forced-air space heaters.

Observations The heating load can be reduced in winter by closing off vents and doors between the grinding/drilling area and the annealing furnace area. Ventilation in the summer requires that they be open.

Waste heat from the annealing furnace can be used to heat the grinding/drilling area during the winter.

The total energy required to heat this space is estimated using the values given in the ASHRAE Standard [5] as a guide.

Heat Loss Through the Building Envelope From Figure 3, page 15, in ASHRAE 90-75, for heating degree days of 3000:

$$U_{ow} = 0.33 \text{ Btu/hr-ft}^2\text{-°F for the outside walls}$$

From Figure 4, page 16:

$$U_{or} = 0.10 \text{ Btu/hr-ft}^2\text{-°F for the roof}$$

From Figure 2, page 14:

$$R = 2.8\text{°F-hr-ft}^2\text{/Btu for the floor (unheated slab)}$$

From the dimensions of the building:

$$\text{Area of outside wall } A_{ow} = 150 \times 18 + 60 \times \frac{18 + 22}{2} = 3900 \text{ ft}^2$$

$$\text{Area of roof } A_r = 150 \times 60 = 9000 \text{ ft}^2$$

$$\text{Area of floor } A_f = 150 \times 60 = 9000 \text{ ft}^2$$

$$\begin{aligned}\text{The total heat loss} &= U_{ow} \times A_w + U_{or}A_r + \frac{A_f}{R} \\ &= (0.33 \times 3900) + (0.10 \times 9000) + \frac{9000}{2.8} \\ &= 5400 \text{ Btu/hr-°F}\end{aligned}$$

Heat Loss Due to Air Infiltration It has been determined by on-site measurement that air flow through the openings in the dividing wall between the grinding/drilling area and the annealing furnace area is as follows:

	Velocity (ft/min)	Area (ft^2)	CFM
Through three ventilators in the wall	55	3 × 6.25	1030
Through a double door (always open)	140	46.6	6520
Total			7550

This air is assumed to enter the working space through the outer door and windows. It must, therefore, in the winter, be heated to the inside temperature. To heat this air requires:

$$\left(7550\ \frac{ft^3}{min}\right)\left(0.0766\ \frac{lb}{ft^3}\right)\left(0.24\ \frac{Btu}{lb\text{-}°F}\right)\left(60\ \frac{min}{hr}\right) = 8300\ Btu/hr\text{-}°F$$

Total Heating Load The total energy required to maintain 70°F inside, with a 40°F outside temperature, is

$$(5400 + 8300)(70 - 40) = 411{,}000\ Btu/hr$$

The energy requirements for the heating of this space in winter can be reduced by $\frac{8300 \times 100}{(8300 + 5400)}$ or 60% if the louvers are closed off and the double doors are rehung and kept closed.

It should be noted that a part of the heating energy entering this space comes through the inside wall from the furnace area.

Recommended Action 1: Close off louvered vents in winter, rehang double doors. This action will reduce the heating load for this area. Since the space heaters are used infrequently (mainly at night) and the temperature in this space floats at other times, the main result will be an increase in comfort level (temperature) rather than a fuel savings. The only cost would be the organizing of the opening and closing of louvers once each year.

Recommended Action 2: Space heat the grinding/drilling area by utilizing recovered waste heat from annealing furnace. This action will again result in an improved comfort level and will also result in the savings of that gas normally used in the space heaters. This amounts to an estimated 55 MBtu per year, with an associated cost of $75 per year which will be saved.

Economic Analysis The implementation of this ECO would be a part of the cost of the total waste heat recovery involving ECOs 1, 2, 3, and 4. Implementation of all of these ECOs would involve a single heat exchanger to recover waste heat from the annealing furnace. The investment versus payback for this is given in Table 7.5. In addition, a payback period analysis is presented for each ECO independently.

The level of capital investment possible for this ECO to obtain payback in a specified number of months at interest rate of 8 and 10% is shown in Table 7.6.

ECO 2: Use of Waste Heat for Drying Oven

Current Practice The drying oven is electrically heated to the modest temperature of 200°F. This heat is rejected at the end of each drying cycle (approximately 20 minutes) by either a purging fan or the opening of the oven doors.

Observations The annealing furnace is always running when the drying oven is in use and waste heat could be used to heat the oven, thus eliminating the need for electrical energy for this purpose.

Table 7.6. Payback Analysis for ECO No. 1

Payback period (months)	Capital investment	
	8% interest	10% interest
6	$ 36	$ 36
12	70	68
18	102	101
24	133	130
30	164	159
36	193	188

Recommended Action Use waste heat from the annealing furnace to heat the drying oven.

Anticipated Savings From an observation of the on/off cycling of the 90-kW oven heater during one complete drying cycle, it was found that the heating element is energized for seven minutes in each half hour.

The energy used per hour in this heating element is

$$2\left(7\ \frac{\text{min}}{\text{hr}}\right)\left(1\ \frac{\text{hr}}{60\ \text{min}}\right)\left(90\ \text{kW}\right) = 21\ \text{kWh/hr}$$

The energy used per month is

$$\left(21\ \frac{\text{kWh}}{\text{hr}}\right)\left(8\ \frac{\text{hr}}{\text{shift}}\right)\left(2\ \frac{\text{shifts}}{\text{day}}\right)\left(5\ \frac{\text{days}}{\text{wk}}\right)\left(4.35\ \frac{\text{wk}}{\text{mo}}\right) = 7308\ \text{kWh/mo}$$

At the January 1977 minimum block rate of 1.212¢/kWh, a savings of $1068 per year could be achieved.

Economics Analysis The implementation of this ECO would be a part of the cost of the total waste heat recovery involving ECOs 1, 2, 3, and 4. Implementation of all of these ECOs would involve a single heat exchanger to recover waste heat from the annealing furnace. The investment versus payback for this is given in Table 7.5. In addition, a payback period analysis is presented for each ECO independently.

The level of capital investment possible for this ECO to obtain payback in a specified number of months at interest rates of 8 and 10% is shown in Table 7.7.

ECO 3: Preheat Combustion Air for Annealing Furnace

Current Practice The hot flue gas from the annealing furnace (measured as 875°F) is exhausted to the atmosphere by roof fans. Some of the energy of the flue gas could be recovered to preheat the combustion air.

Observations Preheating combustion air will improve combustion efficiency, thereby reducing the amount of fuel required to supply a specified heat load. The fuel savings possible by using heated combustion air is as shown in Figure 2.8.

Table 7.7. Payback Analysis for ECO No. 2

Payback period (months)	Capital investment	
	8% interest	10% interest
6	$ 506	$ 506
12	988	964
18	1445	1421
24	1879	1831
30	2313	2240
36	2723	2650

Recommended Action Install a gas to gas heat exchanger to recover heat from the annealing furnace flue gas to preheat the combustion air.

Anticipated Savings The observed fuel flow rate from the rotameter was 1350 ft^3/hr. Assuming operation of the furnace for 18 hours a day, 5 days a week, 52 weeks per year, gives an annual consumption of:

$$\left(1350\ \frac{ft^3}{hr}\right)\left(18\ \frac{hr}{day}\right)\left(5\ \frac{days}{wk}\right)\left(52\ \frac{wk}{yr}\right) = 6.32 \times 10^6\ \frac{ft^3}{yr}$$

Assuming the combustion air is preheated to 600°F will give a fuel savings of 11.5% or 7.27×10^5 ft^3/yr. This savings of fuel will result in cost savings based on the lowest rate (applicable for consumption over 5600 ft^3/mo) of $0.18 per 100. Thus the annual cost savings will be

$$7.27\left(10^5\ \frac{ft^3}{yr}\right)\left(\frac{\$0.0018}{ft^3}\right) = \frac{\$1309}{yr}$$

Economic Analysis The implementation of this ECO would be a part of the cost of the total waste heat recovery involving ECOs 1, 2, 3, and 4. Implementation of all of these ECOs would involve a single heat exchanger to recover waste heat from the annealing furnace. The investment versus payback for this is given in Table 7.5. In addition, a payback period analysis is presented for each ECO independently.

Table 7.8. Payback Analysis for ECO No. 3

Payback period (months)	Capital investment	
	8% interest	10% interest
6	$ 625	$ 625
12	1220	1190
18	1785	1755
24	2320	2261
30	2856	2767
36	3362	3273

The level of capital investment possible for this ECO to obtain payback in a specified number of months at interest rates of 8 and 10% is shown in Table 7.8.

ECO 4: Using Annealing Oven Waste Heat to Heat Hot Water for Drilling and Cleaning Machines

Current Practice The four glass drilling machines use one 12-kWh electric heater each to keep a sump of water at approximately 170°F. These sumps are made of sheet metal with an open top and have dimensions of 3 ft × 2 ft × 15 in. Water is circulated from the sumps to drench the glass drilling operation.

Observations It is necessary to heat the water (which has a cleansing agent added) used in the drilling operation to adequately clean the grease from the glass. Waste heat from the annealing furnace flue gas could be used for this purpose, thus eliminating the use of the electric heaters currently being used.

Recommended Action Use waste heat from the annealing furnace flue gas to heat the water used in the drilling operation.

Anticipated Savings It is assumed that the water in each sump is heated each morning from 60°F to 170°F and maintained at this temperature for two shifts of operation. It is further assumed that the 12-kWh electric heaters operate half the time to maintain the 170°F temperature.

Energy required per sump to heat the water from 60°F to 170°F is

$$(7.5\ \text{ft}^3)\left(6.24\ \frac{\text{lb}}{\text{ft}^3}\right)\left(\frac{1\ \text{Btu}}{\text{lb-°F}}\right)(110°\text{F})\left(\frac{1\ \text{kWh}}{3412\ \text{Btu}}\right) = 15.1\ \text{kWh}$$

Energy required per sump to maintain water temperature is

$$6\ \text{kW} \times \frac{16\ \text{hr}}{\text{day}} = 96\ \frac{\text{kWh}}{\text{day}}$$

Total energy required to operate four units for one year is

$$(4)(15.1 + 96)(5)(52) = 115{,}544\ \frac{\text{kWh}}{\text{yr}}$$

The cost of this energy at $0.014/kWh is $1618 and represents the annual savings possible by using waste heat from the annealing furnace to supply hot water for the drilling operation.

Economic Analysis The implementation of this ECO would be a part of the cost of the total waste heat recovery involving ECOs 1, 2, 3, and 4. Implementation of all of these ECOs would involve a single heat exchanger to recover waste heat from the annealing furnace. The investment versus payback for this is given in Table 7.5. In addition, a payback period analysis is presented for each ECO independently.

The level of capital investment possible for this ECO to obtain payback in a specified number of months at interest rates of 8 and 10% is shown in Table 7.9.

Table 7.9. Payback Analysis for ECO No. 4

Payback period (months)	Capital investment	
	8% interest	10% interest
6	$ 772	$ 772
12	1508	1471
18	2206	2170
24	2868	2795
30	3530	3420
36	4155	4045

ECO 5: Location of Compressor Inlet

Current Practice The air compressor is located in a separate room within the grinding and drilling area (where the compressed air is required). Air intake is inside the compressor unit housing.

The air inside the building is comfortable (estimated 65-70°F) in winter and hot (>90°F) in summer. Furthermore, when the doors of the compressor room are closed air enters that space over the "half-walls." The air supply is thus taken from the upper (hotter) levels of the building space.

Observations Cooler air is generally available outside the compressor room since it is located just inside the northwest corner of the building. The efficiency of an air compressor increases as the temperature of the inlet air to the compressor is reduced.

Recommended Action Supply cooler inlet air to the compressor by installing ductwork to pipe in outside air.

Table 7.10. Effect of Inlet Air Temperature on Air Compressor Efficiency

Temperature of air intake, °F	Intake volume required to deliver 1000 ft^3 of free air at 70°F	% horsepower saving or increase relative to 70°F intake
30	925	7.5 saving
40	943	5.7 saving
50	962	3.8 saving
60	981	1.9 saving
70	1000	0 saving
80	1020	1.9 increase
90	1040	3.8 increase
100	1060	5.7 increase
110	1080	7.6 increase
120	1100	9.5 increase

Source: U.S. Department of Commerce, *Energy Conservation Program Guide for Industry and Commerce*, NBS Handbook 115, U.S. Government Printing Office, Washington, D.C., 1974, p. 3-33.

Anticipated Savings Saving in compressor power as a function of air intake temperature is shown in Table 7.10 which is taken from NBS Handbook 115, *Energy Conservation Program Guide for Industry and Commerce* (EPIC).

Winter Operation Air delivery is at 70°F. Air intake temperature is at 70°F. Air intake could be at 40°F (outside) for a 5.7% saving. This occurs over, say, five months per year (November-March). Assume compressor operates on load for 50% of the time. The potential savings amount to

$$\left(\frac{5.7}{100}\right)(60\ \text{hp})\left(0.746\ \frac{\text{kWh}}{\text{hp}}\right)\left(\frac{50}{100}\right)\left(16\ \frac{\text{hr}}{\text{day}}\right)\left(5\ \frac{\text{days}}{\text{wk}}\right)\left(4.35\ \frac{\text{wk}}{\text{mo}}\right)$$
$$= 444\ \text{kWh/mo}$$

Summer Operation Air delivery is at 85°F. Air intake temperature is at 95°F. Air intake could be at 75°F (outside) for a 3.6% saving. This occurs over, say, five months per year (May-September).

Assume the compressor operates on load for 50% of the time. Then potential savings amount to

$$\left(\frac{3.6}{100}\right)(60\ \text{hp})\left(0.746\ \frac{\text{kWh}}{\text{hp}}\right)\left(\frac{50}{100}\right)\left(16\ \frac{\text{hr}}{\text{day}}\right)\left(5\ \frac{\text{days}}{\text{wk}}\right)\left(4.35\ \frac{\text{wk}}{\text{mo}}\right)$$
$$= 280\ \text{kWh/mo}$$

Annual energy savings = (5 × 444) + (5 × 280) kWh = 3620 kWh.
Annual dollar savings = $44.

Table 7.11. Payback Analysis for ECO No. 5

Payback period (months)	Capital investment	
	8% interest	10% interest
6	$ 21	$ 21
12	41	40
18	60	59
24	78	76
30	96	93
36	113	110

Economic Analysis The level of capital investment possible to obtain payback in a specified number of months at interest rates of 8 and 10% is shown in Table 7.11.

REFERENCES

1. *ASHRAE TRANSACTIONS*, Vol. 82, Pt. 2, 1976.
2. U.S. Department of Commerce, *Energy Conservation Program Guide for Industry and Commerce*, NBS Handbook 115, U.S. Government Printing Office, Washington, D.C., 1974, pp. 3-29.
3. *Ibid*, pp. 3-31.
4. C. T. Horngren, *Cost Accounting--A Managerial Emphasis*, 9th ed., Prentice-Hall, Englewood Cliffs, N.J., 1977.
5. *ASHRAE Standard 90-75*, American Society of Heating, Refrigerating and Air-Conditioning Engineering, Inc., New York, 1975.

Part 3

RESPONSES TO ENERGY CONCERNS

Since the oil embargo in 1973 by the Organization of Petroleum Exporting Countries, many responses have been made by governmental mental units, corporations, and individuals to cope with energy shortages and higher prices.

Industrial corporations use large quantities of energy and the cost of energy is often a significant part of production or transportation costs, so these firms reacted quickly in a variety of ways to the energy "crisis." Several examples of their actions, both operational and strategic in nature, are reviewed in Chapter 8.

The U.S. government also reacted quickly as energy became a national concern. Many existing policies were reviewed and changed and many new programs and organizational units were created. Past policies and post-embargo federal energy programs are discussed in Chapter 9, as well as the implications of higher energy prices.

In Chapter 10, the status of several energy sources is reviewed to provide some perspective on the types of energy that may be utilized during the next few decades. Special emphasis is given to pollution problems associated with each source of energy.

Chapter 8

COMPANY RESPONSES TO ENERGY PROBLEMS

INTRODUCTION

As more and more business firms realize that low cost, plentiful energy will probably never be available again, they have responded in a variety of ways. In *economic* terms, what has happened is that relative prices have changed. Faced with such a change, astute management will alter its use of resources--perhaps by changing the total quantities used and/or altering the proportions in which inputs such as energy and labor are combined. Both in the short and long run there may be a tendency to use less energy. In the long run, capital may be substituted for energy inputs. From the standpoint of *management*, responses may be classified as either *operational* or *strategic* in nature. Operational changes pertain to modifications of policies, procedures, or processes in existing operating systems, with no basic changes in the nature of the business. Strategic changes are those which alter the approach or structure of the firm as it seeks to fulfill its organizational goals. Actual examples of each type of change are reviewed in this chapter.

OPERATIONAL CHANGES

Before changes are made in existing business operations, some motivating force powerful enough to overcome institutional inertia must exist. Such a force was unleashed by the oil embargo of 1973 and by the unparalleled price increases in crude oil which OPEC forced on buyers after they lifted the embargo. These factors, along with the increasing demand for energy throughout the world and the constraints on other types of energy supplies, were sufficient to cause many firms to think hard and fast about how to cope.

The firms which responded first were those where the economic impacts were the greatest. Chemical companies use large quantities of energy in their processing plants as well as hydrocarbon feed stocks for conversion to various chemicals. Thus, they were hit very hard when natural gas and crude oil became scarce and more expensive. In fact, many such companies began to be concerned about energy availability and prices in the late 1960s. The real blow came in the 1970s, however. One large chemical firm, for example, reported that its hydrocarbon price index increased 145% in 1973 (relative to the previous year) and its energy price index increased 73%; the prices charged for the chemicals it produced needed to be 15-20% higher just to recover these costs. Price escalation and shortages continued, and operational impacts intensified.

Operational changes in industrial firms were made for two basic reasons:

1. To reduce the operating costs of energy and hydrocarbon material inputs so the companies could remain competitive
2. To reduce the likelihood of shortages, sudden curtailments or the unavailability of needed energy and feedstock supplies so that operations would not be shut down

Energy Conservation Programs

To formalize the efforts to conserve energy and to better manage the procurement and allocation of energy and related material resources, companies in increasing numbers have started energy conservation programs and have appointed one person or a committee to coordinate the activities [1]. Energy conservation programs, as described in detail in Chapters 5-7, may be established at the corporate, divisional, and plant levels. They offer several advantages over piecemeal efforts by individual managers. First, such a program usually signifies that top management believes energy conservation is important and that everyone should be an active contributor. Second, it provides a central point for coordination of all efforts and for implementing company or unit policies, so that greater efficiency and consistency result. Third, it provides a mechanism for planning new energy conservation programs and policies, and ensures adequate attention to the problems. Finally, it serves as the means to collect quantitative data on the results of energy conservation efforts so that performance can be evaluated in relation to program goals and costs.

Some firms recognized the importance of more efficient energy management even before the oil embargo of 1973. For example, the Texas Division of Dow Chemical, U.S.A., started an energy conservation program in September 1967 [2]. The Olin Corporation also started energy conservation campaigns in 1972, and by the end of 1976 had saved $23.5 million [3]. In the future, more and more firms are expected to initiate such programs.

Price Increases

A common corporate response to higher energy costs has been to increase prices in order to maintain targeted profit margins. Bus and plane fares have been increased to pay for higher fuel costs. Electric utility rates have been increased to pay for higher coal and gas costs, and fuel adjustment clauses have been introduced to make price increases easier and quicker (by avoiding regulatory commission hearings). Fertilizer, an important input in the agriculture sector, and industrial chemicals have increased in price significantly, for both are derived from hydrocarbon feedstocks. The examples of price increases due to higher energy costs are almost unlimited, and such increases are passed on through each stage of the entire production-distribution cycle until the ultimate consumer pays for the higher energy costs. However, in some industries, energy costs are not automatically passed on through higher prices because market factors prevent such action. For example, if a competing firm, either domestic or foreign, has a less costly or more reliable source of energy or related materials, that firm does not need to increase prices (or it needs a smaller increase) to pay operating costs and to maintain profit margins high enough to attract investors and new capital. Also, a competitor with a more modern, energy-efficient plant can hold prices down and often prevent price increases by firms that need one. Thus, the operation of the market economy, which uses the price mechanism to determine resource flows, determines which companies will survive. This is the basic motivation and justification for better management of energy resources, just as for better management of all other resources.

Investments in Equipment or Facilities

In addition to procedural changes in operating systems which required no investment of funds, such as those discussed in Chapter 5, many other changes in operating systems have been made by corporations which required significant levels of investment. For example, typical investments in chemical firms with the highest priority, or greatest energy saving per dollar invested, include: (1) installing combustion controls, (2) replacing obsolete pumps, (3) restoring condensate return systems, (4) revising tower recycle control systems, and (5) repairing or adding insulation [4].

Many investment opportunities to conserve energy involve more energy-efficient equipment or more automatic control devices to reduce energy waste. In addition to better control of industrial processes, computerized systems for control of heating, cooling, and lighting have been installed in several large commercial buildings, offices, and hospitals. Other examples were discussed in Chapter 7.

Scarce metals such as copper and lead have been recycled for many years due to economic benefits to buyers and sellers alike. In more recent years, other materials such as aluminum and glass containers and waste paper have been recycled to reduce solid waste pollution and to provide economic benefits. In the aluminum industry, large companies such as Alcoa and Reynolds have an additional incentive to recycle cans. This process is said to save 95% of the energy it would take to make molten metal from virgin ore [5]. These firms have invested in large recycling facilities and apparently the savings in energy and bauxite and processing costs have made the investments profitable.

Product, Services, and Marketing Changes

Whenever conditions change in the marketplace for any reason, alert business firms always look for new products, modified products, or new services to offer, since there may be an opportunity to meet new demand profitably. Since energy costs have increased dramatically since 1973 and supplies of some types of energy have become unreliable, many changes in products and services have been made. Additions and deletions have occurred due to shifting consumer demands, and some products have been redesigned to make them more energy efficient. Solar collectors have been produced and marketed to households, air conditioners and other appliances have been redesigned to make them more energy efficient, and the home insulation industry has boomed. These are just a few examples of energy market-related responses to higher energy prices.

Process control devices and computerized energy management systems for buildings have enjoyed higher demands as a result of higher energy costs. Also, new contracting services have become available to provide from a single source all interior building construction services--walls, ceilings, air conditioning, heating, sound and light control, and insulation. One firm providing this type of service offers to do the work without profit if it cannot reduce energy costs by 20% and the firm will accept as its profit 50% of the actual energy cost savings during the first two years after the work is completed.

A shift to more fuel-efficient vehicles has occurred and automobile manufacturers are being pushed by the U.S. Congress to increase vehicle fuel efficiency even faster than consumer demand shifts would dictate. Demand is shifting to diesel engines for trucks, since they are as much as 66% more fuel efficient than gasoline engines [6]. Also, the demand for heavier trucks with larger engines has increased in the United States and in Europe because they are more fuel efficient in relation to payload [7]. Truck and automobile producers are modifying their plans to meet these shifts in demand. They have also responded by making sudden shifts in production plans as consumer demand shifted; in late 1973, when the gasoline shortage caused a precipitous decline in large car sales, auto firms converted plants designed to make larger cars so they could produce small cars and they did it faster than industry experts believed possible.

Heat pumps, which are much more efficient than electric furnaces in most situations, are selling at a rapid rate, and all sorts of solar energy devices--many unproven--have reached the marketplace [8].

The product shakeout started in late 1973 due to the fear of an energy crunch and other unfavorable conditions in the business environment [9]. At that time, many companies trimmed their product lines and retained only those with the most profit potential.

In addition to product additions and deletions, many business firms have redesigned existing products to make them more energy efficient. This effort might be viewed as a basic change in business strategy by some firms, but for most of them, it seems to be a modification of existing marketing policies in order to gain consumer loyalty, market share, and higher profits. By changing product characteristics so that operating costs for energy are lower to the consumer, and by promoting this feature through advertisements in magazines, on billboards, and on television, consumers are likely to seek out such products to purchase when they are in the market. This type of design change has been made in automobiles, cookware, heat pumps, truck engines, and air conditioners. Also, the energy-saving nature of other products such as microwave ovens has been promoted heavily in marketing programs.

Eventually, changes in products and services will be made which result in a lower per capita demand for energy as the full costs of energy are reflected in the prices consumers pay. These costs will include those required to obtain the energy, to transport or distribute it, to produce products or provide services, to operate or use the products, and to dispose of the products when no longer useful. Thus, the price mechanism will lead to changes in life-styles, in products and services demanded, and in the ultimate allocation of the resources available to humans.

Employee- and Consumer-Related Programs

Several business corporations, as well as government organizations, have introduced van-pooling programs for their employees in an effort to conserve fuel [10]. The organizations generally purchase vans large enough to haul several employees, obtain a volunteer driver, and solicit riders who pay a fee to the company to cover operating costs and the cost of the van. Benefits include a great reduction in fuel consumed, reduced air pollution, lower transportation costs for the employees, and a reduction in required parking space. The Minnesota Mining and Manufacturing Company reports that it avoided a $2 million investment in parking facilities at its plants as a result of the van-pooling program [11].

Corporations and government agencies have also initiated energy conservation promotion programs for employees and for the general public. They encourage people to reduce energy consumption in their homes, cars, and activities. Bumper stickers, information sheets, brochures with energy-saving tips, television spots, and magazine ads are all used. Some utilities also provide free home inspection service to suggest ways to conserve energy. Home builders are constructing and marketing more energy-efficient houses and the response from home buyers has been favorable [12]. Also, consumers in increasing numbers are realizing that an investment in better insulation can yield a significant return in energy cost savings for many years.

STRATEGIC CHANGES

Significant reductions in energy consumption and resulting cost savings can be realized by operational changes like the preceding examples. However, more fundamental changes in the design of the business or the broad plan used in pursuing goals with limited resources in a competitive environment may lead to even greater savings. Such changes are called *strategic* plans, and may pertain to new types of products, new markets, new organization structures, facility relocations, and new types of integration. They are designed to increase the probability of business goal fulfillment and usually have a major effect on overall corporation financial performance. Several examples of strategic changes which corporations have made in response to the unstable energy environment of the past few years are described in the following paragraphs.

New Energy or Feedstock Supplies

One of the most serious impacts of the changing world of energy for the business manager is the uncertainty about obtaining the amount of energy or feedstocks needed when they are needed. Various firms have experienced a curtailment in the supply of natural gas, electric power, and petroleum products. One large chemical company had its electricity cut off by a utility firm and embarked on a new strategy of backward vertical integration as a result. The plan was to obtain energy and hydrocarbons at their source and the company set out to purchase a coal company to obtain reserves. Utility companies have followed the same strategy [13], and one trucking firm even purchased a gasoline company to ensure a more steady fuel supply. Utilities are increasingly entering into 20-year contracts to obtain coal.

Some companies have drilled for gas near their plant sites to obtain needed supplies. General Motors Corporation drilled and obtained natural gas at its Lordstown, Ohio plant. The Miller Brewing Company obtained mineral leases on 15,000 acres near its Auburn, N.Y., brewery and is spending $20 million to drill wells and to build a pipeline to carry the natural gas to its plants [14].

The disruption of production activities and long periods of downtime due to energy or feedstock shortages can be devastating to an industrial firm. The costs per unit of output increase rapidly and orders cannot be filled on time. These results will seriously reduce profits and can easily lead to permanent plan shutdowns. The problems in foreign operations are sometimes more serious than those in the United States. For example, one General Motors Corporation plant in Iran experienced 15 power failures in 12 months and the plant's 2200 workers were idled each time [15]. Car plants in Iran run at 51% of capacity, partly due to these power outages. "The basic trouble is that energy plants aren't being expanded fast enough to meet the soaring demand" [16]. Thus, the most energy-efficient plants in the world will not perform satisfactorily unless a strategy can be devised to ensure a fairly constant supply of energy.

New Business Ventures

Many business corporations have adopted diversification strategies in order to reduce the risk associated with energy production and distribution or to develop new opportunities which are related to energy. To diversify, a firm changes the design or structure of the business so that new products or services are added and they are usually sold in new markets. Among large petroleum companies, the Mobil Corporation has made a major diversification move by merging with Marcor Inc., a large consumer goods and container company. Mobil is also planning to develop a joint venture coal mine in Wyoming by 1980 and to expand to commercial scale its uranium leaching plant near Laredo, Texas even before then [17]. Mobil's chairman states that these projects signal a company strategy of "placing increasing emphasis on alternative energy sources" [18]. Other petroleum companies are also diversifying into types of energy other than petroleum and natural gas.

A new partnership to construct, own, and operate two tankers for transporting liquefied natural gas to the U.S. from Algeria was planned in the mid-1970s [19]. Panhandle Eastern Pipe Line Company,

General Dynamics Corporation, and Moore-McCormick Bulk Transport Inc. were the companies involved, and the new venture filled a different role in the strategic plans of each firm. In other corporations, similar strategic moves have been undertaken in an effort to avoid threats related to energy supplies and costs or to take advantage of energy-related business opportunities.

Plant Conversions from Gas to Oil or Coal

Petrochemical companies are engaged in major facility construction projects which will enable them to switch from rapidly diminishing natural gas fuel and feedstocks to more plentiful oil [20]. A switch from one input material to another might be viewed as an ordinary operational change, but the scale and investment required in this conversion are so vast that it must be considered of strategic importance. It is said that plants that use oil as feedstock require three or four times the investment for the same amount of output as those that use gas [21]. Thus, the companies involved are trying to avoid a critical shortage of energy in a few years, but they must assume other risks of great magnitude in order to achieve this goal. The return on these large investments may be unsatisfactory, the entire industry may build more capacity than is needed due to uncertainty of demand and of competitors' building plans, oil itself may become much scarcer and more costly, and new technology may be developed to change the economics of various processes and energy types.

Many utilities and industrial firms are also developing plans to switch from gas or oil to coal, since coal appears to be the most abundant source of energy for the next several decades. However, a rapid increase in the supply of coal faces many problems and companies will not convert until the future becomes clearer [22]. Besides the uncertainties about coal supply, the costs of converting many types of old equipment from natural gas to coal are very high. Therefore, many companies seem to be waiting until the shifting U.S. energy policy is stabilized to evaluate the incentives for conversion [23].

Retrenchment Through Demarketing

Many utility companies have adopted a strategy of "demarketing," which means they have stopped promoting the use of more energy-consuming products [24]. The energy crunch of 1973 was a basic reason for adopting such plans; however, some state regulatory agencies, such as the Washington Utilities and Transportation Commission, have banned promotional ads by regulated energy utilities, so the firms have no choice [25]. Some utilities want this type of ban lifted or relaxed [26]. They claim the large reductions in industrial customers could cause them to suffer losses [27]. On the other hand, the large capital requirements for expansion, the uncertainty and expected high cost of future fuel supplies, and the delays associated with more and more regulatory activities and public protests make many utility companies content with a slowdown strategy.

Research, Development, and Exploration Programs

Recent energy-related developments have triggered a wide variety of research and development programs. Many of these have been funded by the U.S. government and nonbusiness institutions, but industrial firms have also invested company funds in such projects. The goals have varied widely, but include the search for ways to convert from one form of fuel to another (coal to gas), for more energy-efficient products, for new or more efficient industrial processes, and for new or improved transportation systems [28].

Petroleum companies have also expanded their search for new oil and gas fields, and are likely to expand even more as energy prices increase.

New Financing Methods

Some novel financial strategies have been adopted by business corporations to obtain funds for exploration and other capital projects. Several petroleum companies have used future production payments as collateral to obtain bank loans [29]. This means that the loans are repaid from a share of the revenue from future production of the property involved. Instead of borrowing the funds directly, the large firms usually sell an interest in future production to a third party, which in turn obtains the bank loan [30].

SUMMARY

Business corporations have responded to changing energy supply and demand conditions in the world in a wide variety of ways. In general, the larger firms have made systematic efforts to reduce energy consumption in all of their operations, and have realized major savings. At the same time, many changes have been made in corporate strategic plans. These include: diversification; vertical integration to obtain energy sources; demarketing; major research, development and exploration programs; plant conversions and expansions; and new financial arrangements.

In the future, similar corporation responses can be expected as firms try to anticipate changes in the business environment. In view of these expectations and each company's strengths and weaknesses, they will develop and implement strategic and operational plans to achieve corporate goals. However, due to the uncertainty and rapid changes expected in the environment, contingency planning will become much more important in the future.

REFERENCES

1. W. A. Rollwage, Energy conservation in chemical plants, *Chemical Engineering Progress*, Vol. 71, No. 10, 1975.
2. *Ibid*, p. 45.
3. *Olin Annual Report--1976*, Olin Corporation, Stamford, CT, 1977, p. 21.

4. W. A. Rollwage, *op.cit.*, p, 46.
5. Recycling works (advertisement by Alcoa), *The Wall Street Journal* Jan. 22, 1976, p. 13.
6. Back to diesel basics at Cummins Engine, *Business Week*, July 25, 1977, p. 150.
7. *Ibid.*
8. *Readers Digest*, Mar. 1976, p. 84.
9. The squeeze on product mix, *Business Week*, Jan. 5, 1974, p. 50.
10. More concerns provide commuting vans to aid their employees--and themselves, *The Wall Street Journal*, July 24, 1975, p. 36.
11. *Ibid.*
12. Saving energy, *House and Home*, Apr. 1976, p. 60.
13. Unit of American Electric buying coal reserves, *The Wall Street Journal*, July 1, 1975, p. 7.
14. How some companies are coping with the energy crunch, *Nation's Business*, Apr., 1977, p. 28.
15. R. Vicker, How a GM man in Iran tackles costs, customs he never saw in U.S., *The Wall Street Journal*, July 12, 1977, p. 1.
16. *Ibid.*
17. Mobil pursues coal mining plan at Wyoming site, *The Wall Street Journal*, June 30, 1977, p. 36.
18. *Ibid.*
19. Three firms plan partnership to build two LNG tankers, *The Wall Street Journal*, Mar. 17, 1976.
20. Petrochemicals: The prodigious cost of facing the future, *Business Week*, July 18, 1977, p. 44.
21. *Ibid.*
22. Increasing use of coal as President proposals faces myriad problems, *The Wall Street Journal*, June 9, 1977, p. 1.
23. Firms put off action on conserving energy pending U.S. program, *The Wall Street Journal*, May 26, 1977, pp. 1, 24.
24. *Bulletin of Atomic Scientists*, Jan. 1976, p. 39.
25. Pacific Northwest utilities cite gas alert, seek to resume promotional advertising, *The Wall Street Journal*, Aug. 3, 1976, p. 42.
26. *Ibid.*
27. *Ibid.*

28. Energy conservation's impact on R&D, *Business Week*, June 27, 1977, pp. 52-58.

29. Strapped for capital, major oil firms sell interests in production to get financing, *The Wall Street Journal*, Jan. 26, 1976, p. 24.

30. *Ibid.*

Chapter 9

GOVERNMENTAL RESPONSES TO ENERGY CONCERNS

THE PAST

Sometimes the best way to understand the present condition and to anticipate the future is to examine the past. On the other hand, the famed athlete Satchel Page cautioned us, in discussing his philosophy of life, to ". . . never look back: something might be gaining on you . . ." Despite that admonition we find it helpful to "look back" at U.S. energy use and governmental policies in order to better understand the historical content of today's situation and, it is to be hoped, to more accurately anticipate the future.

The classical economists, Thomas Malthus and David Ricardo, viewed resource scarcity as the ultimate constraint on the advancement of human society [1]. Likewise, as late as the first decade of this century, British economists feared that an insufficient supply of energy would ultimately slow that country's rate of industrialization [2]. Thus, for many years, the finite nature of specific energy resources such as coal and total resources was recognized and the consequences of running out widely discussed and appreciated.

Toward the end of the 19th century and continuing almost to the present, however, something happened that changed the way economists looked at resources, particularly energy. In the first place, oil in vast amounts was being discovered worldwide. Natural gas, wasted for years, became a plentiful and highly prized fuel with the construction of extensive pipelines in the 1920s. By the late 1950s, there appeared to be so much oil and gas that much of the world felt it did not need its coal mines any more and production fell to all-time lows for the 20th century. Finally, in the early sixties the potential of nuclear energy appeared endless and just around the corner.

A hundred years of rapid expansion on the supply side of the market had made energy a plentiful commodity and, in the United States, at least, it was also relatively inexpensive. In fact, during this period resources in general were viewed less and less as the ultimate limitational factor. Rather, they were to be treated just like other factors of production and used up to the point where their contribution to total product equaled their price. In terms of economics at least, short run concern with efficient resource allocation dominated the longer run concern pertaining to limits to growth.

The incredibly successful industrial development that has contributed to the growth and spread of material affluence in the United States was spurred not only by abundant and varied energy resources, but also by and large by public policies which fostered energy use. Energy was viewed as an essential ingredient in our national growth, not only for release from the bondage of human and animal labor, but also to gain diverse creature comforts and mobility.

Since energy is a vital national resource which affects economic growth, living conditions, and defense posture, the U.S. federal government and state governments have long been involved in shaping the conditions under which energy is produced and marketed. Some government policies have contributed to higher energy prices and reduced supplies; many others have helped make energy cheaper and more abundant. For example, states have regulated production (prorationing) in the name of "conservation," but many have viewed this action as a means to maintain relatively high, stable crude oil prices. Also, when cheap foreign oil economically squeezed domestic producers, protectionist oil import quotas were imposed (on a voluntary basis in 1954 and on a mandatory basis in 1959) and petroleum prices were raised. Sold to the public on the basis that a strong domestic oil industry was necessary to national defense, critics described this policy as the "drain America first" policy [3].

On the other hand, policies such as the percentage depletion allowance, introduced over 50 years ago, gave petroleum companies a tax advantage which increased their profits and encouraged them to increase exploration activities [4]. This policy led to greater energy supplies and probably lower market prices. Other taxation policies, such as the intangible drilling cost allowance, also encouraged corporations to increase their spending for gas and oil exploration. However, these policies were designed to compensate for high exploration risks and to ensure an adequate supply of energy for a growing economy; some argue that exploration and energy supplies would have been greatly reduced without them.

Another government action which affected energy supplies and prices was the Sherman Anti-Trust Act of 1890 which led, among other things, to the breakup of the Standard Oil trust in 1911. This Act sought to maintain a more competitive environment so that consumers could not be "exploited" by a supplier with the power to hold prices at very high levels. Also, in 1938 the government imposed ceiling prices on natural gas intended for interstate sale and has controlled these prices with increasing rigor until very recently. And in 1971, price controls were imposed on domestic crude oil and petroleum products. The purpose of these controls was to prevent "excess" private profits, but the consequences were to lower prices to consumers (and therefore encourage consumption) and to discourage exploration for new gas and oil supplies.

Because of its unique features of long-lived radioactivity and association with weapons, nuclear energy is the most regulated of all energy resources. The Atomic Energy Act of 1946 denied personal ownership, constrained international trade, placed a governmental monopoly on enrichment, and generally put extensive restrictions on private sector activities. Revisions passed in 1954 relaxed controls, especially on technology-sharing with other nations. In recent years the dual role for developing and licensing was separated into two agencies (now the Department of Energy and the Nuclear Regulatory Commission, respectively). Other policies, most notably those embraced in the National Environmental Policies Act, cause the licensing process for nuclear plants to be lengthy, with the result that the commercialization of nuclear power has been slowed.

IMPACTS OF GOVERNMENT POLICIES

Until recent years the net impact of U.S. government policies was to maintain abundant supplies of relatively inexpensive energy which encouraged the substitution of energy for various types of labor and capital. The rapid historic growth of energy demand and utilization in this country can be largely explained by the relative attractiveness of energy prices. The factors which led to our "cheap energy" society were government policies, abundant natural supplies, technological advances, and a private enterprise economic system which channeled investment funds to the most attractive opportunities for profits.

As already noted, prior to the late 1960s, the energy objectives of the U.S. government were (1) to maintain low (i.e., to subsidize) prices for energy resources, (2) to provide maximum availability of energy supplies, and (3) to develop nuclear energy. However, some of the policies which were used to achieve these objectives led to some undesirable consequences, especially when viewed from a long-range perspective. These include the following.

Environmental and Health Impacts

Under traditional cost accounting procedures, the "costs" of an activity which are not directly associated with that activity are ignored. Until recent years, the indirect and delayed costs of environmental pollution which resulted from extracting, processing, and using energy resources were largely neglected. These costs include the damage to living plants and animals and the harm to human health as well as damage to property resulting largely from air and water pollution. These environmental external costs represent part of the difference between private market production and consumption costs and the total cost to society of production and consumption, known as social cost. Furthermore, these external costs of energy production generally have not been associated with the same places or people that have benefited from low priced energy. For example, we now pay about $1 billion a year out of general tax revenues for black lung benefits and will continue to pay large amounts for many years in the future. These costs of underground coal mining were not included in the price of the coal produced in past years. Thus, the benefits of relatively inexpensive coal were enjoyed by one generation and part of the costs were levied on the next generation. As will be discussed further in later paragraphs, public policy developments, especially since 1970, are beginning to rectify these inefficiencies and inequities.

Rapid Resource Depletion

Federal energy policies have encouraged economic growth and rapid natural resource extraction for many years. As of 1977, the amount of oil and gas used by the United States alone in one year required over 100,000 years to accumulate in the whole earth's crust through natural processes. Domestic natural gas supplies, despite the major finds in Alaska, have already peaked out. Oil production is peaking at the present time and the costs of incremental additions to oil and gas supplies are rising rapidly.

Loss of Energy Independence and Balance of Trade

The very high growth rate of energy demand, fostered in part by government policy, began to overwhelm domestic production capability of oil in the United States in the early 1960s. Rather than alter our energy consumption policies or patterns, we turned increasingly to imports as a means to continue the "energy feast." This meant relinquishment of national energy independence, but the possible costs of this did not become really clear until 1973. Each year since then, the United States has become more and more dependent on foreign supplies of energy to satisfy our demands. The influence of this situation on the international political scene is clear. Less transparent but perhaps more urgent, is its impact on the U.S. and international trade balances. The sudden change in the cost of international oil has put pressure on international exchange rates and strained traditional patterns of international trade and investment. Ultimately, the entire structure of world trade will be substantially altered.

SHIFTS IN FEDERAL ENERGY POLICIES

Environmental Protection Movement

The advent of the environmental protection movement in the latter half of the 1960s can now be seen as the first major step toward altering our energy policy (even though it was not viewed that way at the time). The environmental movement arose because of the explosive economic growth and resource demand and the resulting increase in pollution levels in the air, in the water, and on the surface of the earth. A growing population and increasing per capita income and consumption, coupled with a "throwaway" attitude among producers and consumers alike, appeared to exacerbate the situation. Increasing evidence of damage to plants and animals (including humans), to the earth's surface, and to property from ever-increasing pollutants has led to shifts in values and to widespread efforts to reduce pollution levels.

Environmental legislation such as the Clean Air and Clean Water Acts, the National Environmental Policy Act, the Coal Mine Health and Safety Act, and the Occupational Safety and Health Act reflected a realization of major external costs in our economy, and a national decision to bring these costs inside the market accounting system. Since more than half of all environmental problems have been associated with energy production and use, energy industries have been greatly affected by this legislation.

The impacts of the environmental/energy policy decisions were many. In addition to direct impacts on air and water quality, there were at least two important consequences in the area of energy:

1. *Energy prices shifted upward significantly.* We should recall that the actual total social *cost* of energy did not radically change but rather that the external environmental costs of energy production and use were reallocated to the consumers of energy through a higher *price*.

2. *The minimal environmental impacts of natural gas production and use were such that users of other forms of energy, especially coal, began to turn to natural gas.* As a consequence, the demand for natural gas, already high because its price had been held artificially low through price controls, rose to higher levels. Our national store of natural gas fell sharply in the wake of explosive demand growth and the ratio of gas reserves to annual demand fell steadily. The consequence of this response was that we enjoyed a rapid move toward improved environmental quality but at a cost of perhaps an inefficient rate of depletion of natural gas reserves.

The Energy Crisis of 1973

The second major shift in our national energy policy did not arise indigenously like the environmental movement. Instead, it was externally imposed by the oil embargo placed on shipments to the United States by the Arab states during the Arab-Israeli conflict of 1973-74. In retrospect, that traumatic event was important and served as effective shock therapy for the nation. Adlai Stevenson once observed that "man does not seem to see the handwriting on the wall until his back is up against it." The handwriting on the wall about the hidden costs of becoming dependent on foreign oil was made very clear as a result of the embargo. It drove home a realization of the finiteness of oil and gas resources and also of the extent of our growing dependence on nondomestic supplies. Perhaps more important for the short run, it provided a real specter of the uncertainty of availability of energy and triggered changes in consumer buying decisions. The embargo crystalized a national concensus about the need to develop domestic energy resources and the imperative to increase the efficiency of energy utilization in the United States.

Although the embargo was short-lived, it clearly established the near-monopoly position held by a very few oil-producing countries. On January 1, 1974, OPEC chose to exercise its power by more than doubling the world market price of oil from about $5 per barrel to about $12 per barrel.

For countries that were heavily dependent on imported oil, these events did define a clear-cut and immediate crisis. In less than three months, balance of payments positions were thrown into disequilibrium, industrial development plans became subject to reconsideration, and the threat of future supply disruptions loomed continuously on the horizon. Virtually every country was affected, especially (1) the industrialized nations of Western Europe, (2) Japan, and (3) the Third World nations. In the last case, the situation was most intolerable; for these less developed countries higher oil prices meant a deterioration in an already low standard of living. Western Europe and Japan theoretically could better afford higher prices. However, with their vulnerability clearly exposed, they became subject to political forces unforeseen but a few months before, for it was quite apparent that their economies and standards of living were inexorably tied to the availability of OPEC oil.

Western Europe and Japan imported almost all the oil they consumed in 1973. The United States, on the other hand, imported less than one-third of its oil at that time. This fact made the situation facing the United States categorically different from that of the other western industrialized nations. Once the embargo ended, lines at gas stations tended to disappear and the crisis atmosphere dissipated rapidly. The situation facing the United States was not one of short run disaster but rather an invisible energy crisis of long run consequence that would affect the population some time in the future.

Perhaps the greatest significance and benefit to the western world of the Arab embargo and subsequent OPEC price increase was to shock it into reconsidering the conventional wisdom on energy. In examining the dilemma in which they now found themselves, leaders of many nations discovered that the seeds of the crisis had sprouted years before and many prophets had appeared, only to see their warnings go unheeded. For example, in the late 1950s, M. King Hubbert predicted that U.S. oil production would probably peak out about 1970, but apparently nobody was too concerned [5]. Textbooks on industrial organization regularly portrayed the coal industry as the archetypical "sick" industry but that was little cause for alarm as late as 1972 [6]. A technical monograph which sought to reintroduce resource scarcity and limitation into the mainstream of economics appeared in 1963. In the preface, Joseph L. Fisher, then President of Resources for the Future, appeared to be almost apologizing for the book when he wrote:

> No problem is of deeper concern to mankind than that of coming to terms with the natural environment so that it will support a reasonably satisfying way of life. With the spread of modern science and technology during the past century and a half the problem has changed somewhat, at least for the economically more advanced areas of the world where the pressure of population on the resource base has been reduced. In such countries, and pre-eminently in the United States, there is no present threat of general resource scarcity--indeed, for basic agricultural products there is glut--but the age-old goal of winning a higher level of living from the soil, the water, and the subsoil minerals continues to call forth the enterprise

> and labor of large numbers of people as well as to tax, at times, the ingenuity of policy-makers. Even though technological innovation and managerial skill appear to assure Americans of adequate quantities of raw materials, the prospect of qualitative deterioration of the environment--in its livability and its aesthetic appeal--must concern us as a problem of the first rank [7].

The impacts caused directly and indirectly by the oil embargo were extensive. Probably the most dramatic effect was the rapid shift toward more efficient automobiles. The average performance (miles per gallon) of General Motors' new car sales increased by over 30% within two years. Legislation was passed to require the sales-weighted "average" new car produced in 1985 to have twice the mileage performance of the "average" 1973 car. It now appears probable that total gasoline consumed will remain approximately constant until the mid-1980s, even assuming a continued increase in the total number of vehicles and vehicle-miles traveled.

The second major impact following the Arab embargo was the discovery of extensive waste in the U.S. energy system. Consumption and production patterns were geared to a low energy price economy. The post-embargo energy market called for a reevaluation of the ratio of energy input per unit of output. For example, we found the illumination levels had increased dramatically and that the nation was grossly "overlit." About 20% of all electricity generation was being consumed in lighting, but roughly half of the electricity being used in lighting was not needed to improve visibility and working conditions. The amount of energy wasted in lighting alone was equivalent to burning 150 million tons of coal per year, if one includes the extra air conditioning required to cool buildings in the summer that are overheated by waste energy from lighting. In the post-1973 period, the myth that it isn't practical to save a significant amount of energy has been dispelled. Also, in examining our energy consumption patterns, the myth that greater energy consumption always leads to greater productivity and economic progress has begun to tarnish.

Post-embargo Federal Energy Programs

Office of Energy Conservation

During the early months of 1973, U.S. energy planners in the Congress and in the Administration faced possible gasoline shortages in the coming summer and, despite large petroleum imports, shortages of heating oil in the following winter. The Office of Energy Conservation was created in the Department of Interior (1) to encourage greater public awareness of the need for energy conservation and (2) to coordinate a Presidentially established program to reduce energy consumption by Federal agencies 6% below projected demand within a year. During the summer and early fall of 1973, the Atomic Energy Commission, in response to a Presidential request, outlined a program for expanded energy research and development and included for the first time proposals for R&D on energy utilization. These relatively quiet stirrings were tossed into the maelstrom of the Arab oil embargo and OPEC cartel price change in late 1973, and emerged as fledgeling federal programs in energy management.

Project Independence

The initial reaction to the onset of the energy crisis in the United States may be characterized as a combination of bravado and disbelief. The bravado found its ultimate expression in Project Independence, which called for a rapid U.S. transition to domestic energy self-sufficiency. Disbelief found its expression through amazement that the world's most powerful industrial country could somehow be at the mercy of a group of nations traditionally relegated to Third World status, and betrayal in the sense that somehow U.S. defenses had been allowed to decay to a point close to the jugular. In the years following 1973, bravado has given way to sagacity; however, many Americans still doubt the very existence of an energy problem.

Project Independence can be interpreted in several ways. Some may say it was a typical American response to adversity; that is, set out to defeat it at whatever the cost. Alternatively, it can be viewed as a pep talk from the coach during a dark hour. In retrospect, however, probably the best analogy is that of a hurt child proceeding on instinct rather than on reason and facts.

The principal goal of Project Independence--energy independence by 1980--was a totally unrealistic objective. To have believed otherwise indicated a lack of understanding of two key facts of the U.S. energy situation: (1) rapidly increasing demand and (2) declining domestic production. To turn such a situation around in less than a decade without intolerable costs was an impossible dream.

FEA and ERDA

In the face of the Arab embargo, President Nixon established (by Executive Order), in October 1973, the Federal Energy Office to orchestrate, in cooperation with the states, the U.S. response to the embargo. This Office was transformed into the Federal Energy Administration (FEA) in 1974. FEA dealt mostly with energy policy, contingency planning, and energy regulatory matters. Also during 1974, the Atomic Energy Commission, together with portions of the Department of Interior and Environmental Protection Agency, were transformed into two new agencies: the Nuclear Regulatory Commission and the Energy Research and Development Administration. The latter was given broad R&D responsibilities in all phases of energy production and utilization.

The federal agency energy reduction campaign with the goal of a 6% reduction in one year was transformed into the Federal Energy Management Program (FEMP). Within three years, a reduction of nearly 30% of projected energy consumption was achieved by federal agencies. FEMP's successes, largely achieved by copying successful programs of selected private industries, were doubly important. First, significant amounts of energy and money were saved; second, a documented, successful model program was completed which served as a national example to prove that major energy savings are possible through careful management practices without significant deleterious impacts on organization performance.

Energy Policy and Conservation Act

After the termination of the Arab oil embargo, the attention of both citizens and policy makers turned mostly to other matters. Higher energy prices began to affect some usage patterns and purchases, but very few important new public policy decisions were made until the passage by the Congress in December 1975 of the Energy Policy and Conservation Act (EPCA, PL94-163). This Act, the most important single policy initiative taken by the United States since 1973

1. Grants specific standby authority to the President, subject to congressional review, to impose rationing, to reduce demand for energy through the implementation of energy conservation plans, and to fulfill obligations of the United States under the international energy program
2. Provides for the creation of a Strategic Petroleum Reserve capable of reducing the impact of severe energy supply interruptions
3. Increases the supply of fossil fuels in the United States, through price incentives and production requirements
4. Conserves energy supplies through energy conservation programs, and where necessary, the regulation of certain energy uses
5. Requires improved energy efficiency of motor vehicles, major appliances, and certain other consumer products
6. Reduces the demand for petroleum products and natural gas through programs designed to provide greater availability and use of our coal resources
7. Provides a means for verification of energy data to assure the data's reliability

Four of the seven objectives of EPCA are concerned with energy conservation. The Act recognizes the vital roles appropriately played by state government. It encourages each state to develop and implement State Energy Conservation Plans (SECP) to reduce the rate of growth of energy demand. Each state was invited to submit a plan which sets forth a goal by that state to reduce its 1980 energy consumption to a level at least 5% below the current projections of 1980 consumption.

Energy Conservation and Production Act

In August 1976, Congress passed the Energy Conservation and Production Act (ECPA, PL94-385). ECPA represents an extension and

and augmentation of many of the provisions and programs identified in EPCA. The stated purposes of this Act are

1. To amend the Federal Energy Administration Act of 1974 to extend the duration of authorities under such Act
2. To provide an incentive for domestic production
3. To provide for electric utility rate design initiatives
4. To provide for energy conservation standards for new buildings
5. To provide for energy conservation assistance for existing buildings and industrial plants, and for other purposes

Department of Energy and the National Energy Act

Energy-related responsibilities that were distributed over several agencies (including Federal Energy Administration, Energy Research and Development Administration, Federal Power Commission, and portions of the Department of Interior) were combined in 1977 into a new cabinet-level agency, the Department of Energy (DOE). This organization was designed to provide better coordination between research and action programs, and a more comprehensive approach to solving our overall national energy problems.

The National Energy Act, passed by the Congress in 1978, started as a DOE effort to develop a comprehensive energy program. The final bill was modified considerably from the original proposal of President Carter during many months of debate by Congress. It was really a composite of five separate bills, as follows:

1. The National Energy Conservation Policy Act of 1978
2. The Power Plant and Industrial Fuel Use Act of 1978 (Coal Conversion Act)
3. The Public Utility Regulatory Policies Act of 1978
4. The Natural Gas Policy Act of 1978
5. The Energy Tax Act of 1978

The National Energy Conservation Policy Act (PL95-619) provides for:

1. An energy conservation program for residences to be offered by utilities
2. Weatherization grants for low income families
3. A solar energy loan program
4. An energy conservation loan program for home improvements
5. Grants to improve energy efficiency in schools and hospitals
6. Energy audits for public buildings

7. Energy efficiency standards for major home appliances
8. Civil penalties on automobile manufacturers who fail to meet fuel economy standards established by EPCA

The Power Plant and Industrial Fuel Use Act (PL95-620) includes:

1. Prohibition against use of oil or natural gas in new industrial and utility plants
2. Restrictions on existing coal-capable large boilers
3. Restrictions on users of natural gas for boiler fuel
4. A pollution control loan program for utilities
5. Supplementary authority to prohibit use of natural gas in other heating and lighting applications

The Public Utility Regulatory Policies Act (PL95-617) provides for:

1. Eleven voluntary rate design standards
2. A requirement for state regulatory authorities to consider each standard within stated time limits
3. Consideration by gas utilities of standards for service termination procedures and advertising expenditures
4. New rules favoring industrial cogeneration facilities
5. Greater federal control of electrical transmission and rates between generating systems
6. Aid to help states implement new requirements and to permit consumer intervention in proceedings
7. A loan program to aid development of small hydroelectric projects
8. Regulations to expedite crude oil transportation within the United States

The Natural Gas Policy Act (PL95-621) provides for.

1. Maximum prices for various categories of natural gas
2. Deregulation of prices of new gas and certain interstate gas as of January 1, 1985
3. Incremental gas pricing so that industry is charged more than residential users
4. Emergency authority for the President to permit gas sales to high-priority users in case of a shortage

The Energy Tax Act (PL95-618) provides for:

1. Residential insulation and conservation tax credits
2. Residential solar or wind equipment tax credit
3. Exemption of gasohol from excise tax
4. A graduated excise tax on gas-guzzling cars
5. Tax incentives for development of geothermal resources
6. Extension of favorable minimum tax treatment of intangible drilling costs for oil and gas

7. Business tax credits for industrial investment in alternative energy property (for example, boilers for coal)
8. Denial of investment tax credit and accelerated depreciation for new gas and oil boilers [8]

The broad scope of the provisions of the Natural Energy Act of 1978 indicates that energy issues are vast and complex. They are also shifting rapidly, as changes occur in the social, political, economic, technological, and physical segments of the total environment in which we live. Therefore, we should not expect that national energy policy will be stable, but instead will be one that continually changes by a process of accommodation to shifting priorities and conditions.

IMPLICATIONS OF HIGHER ENERGY PRICES

The conditions that led to an era of low and falling energy prices must be considered history. As a result of the escalating energy price structure, many different implications are evident.

1. Our energy supplies, heretofore primarily oil and gas and some coal, must be expanded to include other energy sources, such as geothermal and solar, which were previously uneconomical. As the relative price of energy increases, the type and number of "new" energy sources that become economically attractive increase sharply. *The fastest way to discourage investment in a more diversified energy supply system would be to roll back, via price controls, present prices to previous levels, or even to hint that this might be done as a matter of national policy.*

2. Higher energy prices make it economically attractive to alter our energy consumption infrastructure to one of much higher efficiency. This can be effected through direct substitution (e.g., better insulation instead of bigger furnaces and air conditioners), increased efficiency of conversion (e.g., more efficient air conditioners and automobiles), and minor modifications in life-style (e.g., set back thermostats at night, use returnable beverage containers, plan automobile trips more carefully). In almost every instance, such efforts to increase efficiency also increase domestic

economic activity and improve environmental quality. The evolving federal energy policy calls for various specific proposals that provide incentives for investments in conservation, and encourages market pricing regulations and standards to require more efficient energy use.

3. The new energy price level means that it is now important to reevaluate the benefit/cost ratios of various public policies that were created during an era of relatively low energy prices. In the public sector, this includes policies on freight rates and carrier regulation, the oil depletion allowance, and power plant cooling requirements. All of these policies may not need to be changed, as energy prices may have very little effect on their basic validity. However, if public policies such as the ones on environmental protection are based on benefit/cost relationships, then all policies should be reexamined whenever significant changes occur in either the benefit or cost variables.

4. With higher energy prices, external costs of production are a less significant part of total costs. Several years ago, the price of coal was such that the environmental protection costs were estimated to be close to 50% of its market price. With higher market prices for energy, the business of internalizing external environmental costs can be carried on with much less potential disruption of the marketplace.

The points made in the preceding discussion bring both good news and bad news. On the positive side, the substitution of materials, time, labor, and human ingenuity for energy, simply because it is cheaper to do so, also benefits the environment and the economy. Likewise, the nascent shift away from dependence on fossil energy resources toward the use of long-term sustainable energy sources is accelerating and this is desirable, for it may require a half-century or more to complete the shift.

On the negative side, there are near-term economic dislocations caused by such a sudden change of events. Also, a major long-range concern is the implied massive shift over the coming decades toward increased coal production and combustion in response to higher priced and more uncertain quantities of oil and gas. If we use present technologies for mining, conversion, and combustion of coal, the environmental implications are very serious. As reforms in mine reclamation are implemented and advanced combustion (fluidized bed combustion) and pollution control technologies are developed and used, we will have overcome some major problems. Yet even after these steps, the inescapable combustion product, carbon dioxide, could still severely limit the rate at which coal can be used without causing unacceptable planetary climatic problems. This issue, only recently recognized, has profound implications for any national policy which depends upon major expansion of coal production.

REFERENCES

1. T. R. Malthus, *Principles of Political Economy*, Augustus M. Kelley, New York, 1951; also David Ricardo, *Principles of Political Economy and Taxation*, George Bell, London, 1908.
2. W. Jevons, *The Coal Question*, 3rd ed. rev., Macmillan, New York. 1906.
3. E. H. Shaffer, *The Oil Import Program of the United States*, Praeger, New York, 1968; also U.S. Cabinet Task Force, *The Oil Import Question*, U.S. Government Printing Office, Washington, D.C., 1970.
4. S. L. McDonald, *Federal Tax Treatment of Income from Oil and Gas*, Brookings Institution, Washington, D.C., 1963.
5. M. K. Hubbert, *Energy Resources*, Publication 1000-D, National Academy of Sciences-National Research Council, 1962.
6. W. Adams, *The Structure of American Industry*, 3rd ed., Macmillan, New York, 1961.
7. H. J. Barnett and C. Morse, *Scarcity and Growth*, Johns Hopkins Press, Baltimore, Md., 1963.
8. *The National Energy Act*, U.S. Department of Energy, 1978.

SUGGESTED READING

S. D. Freeman, *Energy: The New Era*, Walker, New York, 1974.

D. E. Kash, et al., *Our Energy Future*, University of Oklahoma Press, Norman, 1976.

J. C. Sawhill, ed., *Energy Conservation and Public Policy*, Prentice-Hall, Englewood Cliffs, N.J., 1979.

Federal Energy Administration, *Project Independence*, U.S. Government Printing Office, Washington, D.C., 1974.

U.S. Congress, Office of Technology Assessment, *An Analysis of the ERDA Plan and Program*, U.S. Government Printing Office, Washington, D.C., Oct. 1975.

U.S. Congress, Office of Technology Assessment, *Analysis of the Proposed National Energy Plan*, U.S. Government Printing Office, Washington, D.C., Aug. 1977.

Chapter 10

TECHNOLOGY, ENERGY, AND ENVIRONMENTAL PROTECTION

INTRODUCTION

The explanation in Chapter 1 of the creation of our solar system indicates that existing energy resources of various forms originated from the sun or preceding stellar matter. These resources are an integral part of our physical environment on earth, but as we extract, transport, convert, and utilize them, we often damage other parts of the environment. Heightened awareness of these environmental problems, as well as the depletion of previously abundant energy resources, has led to programs and policies which move toward recognizing the full private and social costs of energy, both as we use it and as we contemplate the supply configuration for the future.

Advances in technology have produced the life-style enjoyed by people in developed nations today, but they have also led to a very energy-intensive form of civilization. In the future, research and development activities are expected to continue at a high level in pursuit of new technologies to help solve energy-related problems. Specifically, advances in technology will be sought to:

1. Find and extract untapped energy resources and to convert economically from one form of energy to another (energy supply technologies)
2. Utilize various forms of energy more efficiently (energy utilization technologies)
3. Control pollutants which arise from energy production and utilization (pollution control technologies)

In this chapter, various forms of energy which are presently available or under development will be discussed. The focus is on problems related to the production and utilization of each form of energy, with special emphasis on pollution control problems. Throughout the chapter, the critical role of technology in providing new sources of energy, in making existing sources more economical and more abundant, and in controlling pollution will be highlighted. Also, a historical perspective of energy supply and utilization shifts is provided to remind us that societies have faced supply problems with particular energy sources before, yet managed to develop new sources and to modify their energy usage patterns to alleviate the problem.

OIL AND GAS

Petroleum and natural gas are the major energy resources used in the United States and in many other nations today. Oil, like coal, is a "leftover" from ancient vegetable growth. Its chemical contents change from one oil field to another. Like coal, the most important "contaminant" in oil is sulfur. Unlike coal, however, the sulfur in crude oil can be removed (during the refining process), albeit at a cost. For this reason "sweet" crude (low sulfur) brings a premium price. Oil also contains heavy metals (such as vanadium) that present a potential environmental problem whose magnitude has not yet been carefully assessed.

After gasoline, diesel oil, and other fuel and petrochemical products have been refined from crude, the residual ("bottom of the barrel") is used for electricity production and asphalt. When residual oil is burned at a power plant or in industrial boilers, air pollution is created by effluents which contain oxides of sulfur and nitrogen. However, residual oil causes only one-third as much air pollution as coal!

The origin and production of natural gas are closely associated with oil. Both fuels are often taken from the same well. Almost identical environmental problems are encountered in extracting natural gas, but it requires little processing before use. Natural gas is transported in underground pipes to points of use, so there is little environmental impact before combustion. It is by far the least environmentally damaging of the several fossil fuel aternatives. Total air pollution is less than 5% of the emissions from coal combustion (per Btu), and there are no solid wastes to contend with. The only impact on water resources is waste heat produced through condensor cooling water.

COAL

During the 1800s, coal replaced wood as the United States' principal energy resource, just as petroleum replaced whale oil for special applications. By the turn of the century, coal accounted for over 80% of total U.S. energy demand. Annual production, all from underground mines, amounted to 212 million tons. During the 20th century, coal gradually gave up its leadership to oil and gas. Its contribution to total demand dropped from 88% in 1900 to 38% by 1950 and 20% by 1970. Production peaked (630 million tons) in 1947 and then declined substantially. Coal lost its dominance because petroleum and natural gas were price-competitive, cleaner, and more convenient to use. In recent years there has been a resurgence of demand for coal and production has exceeded the 1947 peak. As a percentage of the total energy supply, however, coal will never again approach its earlier dominance of the market.

The Changing Market

After World War II ended, the technology of large diesel engines became so successful that coal-fired steam locomotives quickly disappeared, thrusting the coal industry into another round of deepening depression. Government policies, including the oil depletion allowance and interstate natural gas price regulation, helped ensure the unhealthy condition of the coal industry. The market for coal became steadily more dependent on purchases by electric utilities, which rose from 47% to 63% of coal sales between 1965 and 1973.

Demand from that sector increased at a rapid rate, offsetting the continued erosion of other coal markets. At the same time, the federal policy of moving our electrical economy toward dependence on nuclear power (and accompanied by reiterated promises of how cheap it would be and how soon it would arrive) cast a pall of doubt over the future demand for coal. This resulted in a strong financial institutional restraint on long-term investments in coal production. Coal was dirty and had many seemingly intractable problems of both mining and burning. Strip mining (and augering), "cheaper" but more environmentally damaging than deep mining, increased from about 10% of all U.S. coal mined in 1940 to 30% in 1968 and 50% in the early 1970s. Coal, by a large measure our greatest fossil fuel resource, became the ugly stepchild of America's energy family.

During the 1960s, the market price for coal came under further downward pressure due to competition from relatively inexpensive foreign oil. Profitability was marginal and stiff resistance met any actions which tended to increase coal price since demand was price elastic due to interfuel competition. Thus, when environmental concern led to health and safety standards and land reclamation requirements, the coal industry put up great resistance because it was rapidly losing to clean, cheap natural gas as well as to the competition from foreign crude oil.

As the decade of the seventies appeared, electrical demand maintained its rapid growth. The much promised nuclear option was tarnished by escalating costs, time delays, and new questions about safety and security. Natural gas availability became very questionable in the long run because its price was regulated at a level too low to encourage exploration and because the easy-to-get reserves were being depleted. Consequently, coal demand projections took on a more rosy glow. Upward cost pressures in underground mining (e.g., health and safety standards) were offset by a higher labor productivity and lower costs in surface mining. Thus, a long-term upward demand trend for coal was cautiously projected.

Mining

Despite increased attention to safety, the mining toll of human life has been high. Over the past 20 years, more than 6000 lives have been lost in underground mining accidents and many thousands more have suffered injuries and shortening of life spans due to black lung disease. Fatal accident rates in underground mines have fallen slowly over the past 20 years and are now at the level of one fatal accident per 2 million tons of coal produced. Large mines generally have much better safety records than small ones. Since the man-hours required per ton of coal produced are much less for surface mining, the rate of fatal accidents *per ton of production* is one-half to one-third as high for surface mining as for deep mining. Also, injury frequency rates are less by about a factor of four. The hope for the future is that more sophisticated underground mining technologies and implementation of more stringent health and safety regulations can virtually eliminate black lung and cut accident rates to the levels associated with surface mining.

Utilization

When coal is burned in a modern boiler (a 1000-MW power plant will burn about 10,000 tons per day) there are two principal problems: material emissions (sulfur and nitrogen oxides, ash particles, etc.) and waste heat emission (less than 40% is converted to electricity, about 15% goes up the stack, and the rest is discarded in a body of water or a cooling tower or both). The heat emissions are not directly harmful to humans, but can cause major damage to aquatic life. During the early 1970s, rather stringent requirements were placed on thermal discharges.

Atmospheric emissions from coal combustion present a major problem for humans as well as for other animals and for plants. Coal contains 10 to 20% ash, 1 to 5% sulfur, and various other elements, including toxic heavy elements such as mercury, cadmium, and radium. The kind and amount of impurities in coal reflect the conditions under which it was formed hundreds of millions of years ago.

Generally speaking, the lowest sulfur-containing coal contains the highest amount of ash (western coal). Midwestern coal (such as that from Indiana and Illinois) usually contains much sulfur (up to 5%). Appalachian coal varies in impurities, but generally has medium quantities of sulfur (1 to 3%) and ash (5 to 10%).

A 1000-MW plant burning coal with 10% ash, 2% sulfur, and 1 ppm mercury will produce each day about 1000 tons of ash, 250 tons of sulfur dioxide, and 20 pounds of mercury. Precipitators will remove 90 to 99% of the fly ash (most of the large particles but only half of the smaller particles, which are much more important in health problems), but won't remove the sulfur dioxide, mercury, and other materials. These pollutants are presently dispersed in the atmosphere through tall stacks. About 80% of all atmospheric emissions of sulfur oxides are produced in power plants and industrial boilers. The exact impact on human health of sulfur oxides, especially when accompanied by particulates, is not very well known. For example, sulfur dioxide (SO_2) by itself is not a serious problem at low concentrations. However, chemical transformations occur in the atmosphere and in human lungs in which SO_2 becomes further oxidized to sulfuric acid or sulfate salts. These compounds are a major factor in lung disease, now endemic in parts of the country characterized by high air pollution. Sulfur emissions can in some instances improve soil productivity since sulfur is an important element in plant growth. At the same time, the sulfur can cause "acid rain" and adversely affect both soils and aquatic life.

It is ironic that while fly ash precipitators can be made to work relatively well when using high-sulfur-content coal, they work much less satisfactorily with low-sulfur coal. Although precipitators do not trap the sulfur, its presence enables the fly ash to be more readily trapped. When utilities switched to lower-sulfur coal to cut sulfur emissions, precipitator performance dropped severely (for instance, fly ash emissions increased by an order of magnitude).

What can be done? Coal is our most abundant fossil fuel. How can it be mined with less adverse environmental impact? Surface mining know-how is on hand to reduce damages vastly under current practice. For eastern coal, the extra cost will vary but will be no more than several dollars per ton, which ultimately translates to perhaps a 2% increase in the cost of coal-derived electricity (and a vastly improved environment in Appalachia). In the West the situation is different, but there is both good news and bad news. The good news is that western coal seams are very thick (100 ft is not uncommon) and close to the relatively level surface. Thus, reclamation costs (per ton) are much lower than in the East. However, the bad news is that underground water frequently moves through the coal seam and the rainfall is so scarce that a minor surface disturbance can destroy range land. Therefore, it is essential to develop plans and gain commitments in advance to restore the land to productivity. Despite these problems, we can essentially eliminate the worst problems associated with surface mining if we accept the slightly higher prices.

The most knotty problem about coal is how to burn it with less environmental impact. There are several possible solutions:

1. Burn it in remote areas where the emissions "won't hurt anybody."
2. Build tall stacks or use intermittent controls ("the solution to pollution is dilution").
3. Trap particles and sulfur from the exhaust gas.
4. Modify the combustion process so that the particles and gases are trapped right in the boiler.
5. Treat the coal before burning by turning it into gas or by reducing its sulfur content.
6. Increase the efficiency of the combustion process so that more of the heat is used and less must be dumped as waste heat.
7. Use cooling towers and other devices to minimize environmental damage from heat dumps into water bodies.

Let us respond briefly to each of these points:

1. Who has a remote area available these days?
2. Nature has a remarkable capacity to absorb environmental abuses. That capacity is a resource worth using, but carefully; we've clearly overdone it lately. Tall stack strategies may give local relief but spread the trouble to many more people, albeit at a lower level of concentration.

3. We're now trapping particles at almost every plant. The technology for sulfur "scrubbing" is still not fully satisfactory and needs further development.
4. One idea, called "fluidized bed combustion," is about ready for large-scale pilot plant testing. In it the ash and sulfur are largely trapped within the combustion chambers. This process cannot be counted on to give much relief to problems at existing plants but may provide a great deal of help in new plants built several years from now.
5. Coal can be converted into a gas or liquid and then burned fairly cleanly. However, the conversion process is costly and has its own environmental problems. For one thing, nearly half of the energy inherently contained in coal is lost in the conversion process. Alternatively, coal can be "beneficiated" by mechanical and chemical means which can remove at least half of its sulfur at little energy cost. Both of these approaches merit accelerated development.
6. Thermodynamically, the efficiency of coal-fired boilers depends upon how hot they can get compared with the cooling water. In a steam generator, the limits of strength of materials are reached (due to high temperature and high steam pressure) at a level corresponding to an electric generating efficiency of about 40%. If the boiler is made more complicated, for example, by using a gas turbine "topping cycle" followed by a conventional steam turbine, the overall efficiency can increase to 50% or more. The potential environmental benefit of increasing conversion efficiency is twofold: less waste heat is generated and less fuel resource is required per unit of useful power produced.
7. Cooling towers and other ways to lessen the burden of waste heat on the aquatic environment have recently come into wide practice. In some places (such as Biscayne Bay in Florida, water is naturally so hot in summer that even small amounts of waste heat can cause great damage. However, in other places, direct discharges of waste heat can have little effect on aquatic life, and much of that beneficial. Therefore, we must be careful to make site-specific judgments on what to do with waste heat.

In summary, coal production and combustion are our worst polluters when uncontrolled, but can be regulated to a large extent. If current environmental control technology is applied to the coal system electricity (derived from coal), cost will increase about 30%. Most (about 80%) of that increase would be due to air pollution control.

NUCLEAR ENERGY

Decades of research following the discovery of radioactivity revealed two ways in which energy might be derived from the nuclei of atoms. First, a very limited number of very heavy nuclei release a great deal of energy if they are broken up (fissioned) into two or more smaller nuclei. Second, a very limited number of very light nuclei give up a great deal of energy if used together (fusion). It has been only during the past 40 years that we have begun to understand these processes and from them to solve the mystery of how the sun and other stars release so much energy.

The catalyst for fusion is a combination of extreme temperature, pressure, and density. So far we don't know how to control such conditions on earth. However, about 30 years ago we did learn how to control fission--to make it happen as slowly or as fast as we wished. After a quarter century of hard work, energy from fission has become a commercial reality which is economically competitive with alternative sources. Nuclear energy and coal are viewed as the major fuels for providing electricity, especially over the next half century. Nuclear plants presently account for 10% of overall U.S. electricity production. Historically high demand growth rates, if continued, would require very fast construction of new plants. Fortunately, the new electricity demand projections indicate a much lower growth (perhaps 4% rather than nearly 8%) than occurred in 1960-1970. Even at the lower growth rate we will require about 16 new 1000-MW plants every year, each costing well over one-half billion dollars.

Uranium Mining and Processing

The earth has a very limited supply of high grade uranium ore--so limited that within several decades uranium fuel costs can be expected to rise appreciably as we mine lower and lower grade ores. Although fuel costs are much less important for a nuclear plant than for a coal-fired plant, sharply increased uranium costs are not unlikely if demand growth projections hold true. One way to minimize that cost escalation is to design nuclear power plants so carefully that sufficient "surplus" neutrons are produced to convert some of the much more plentiful but nonfissioning uranium (U-238) into fissionable plutonium (Pu-239). If the reactor is carefully designed and constructed, slightly more plutonium is produced than fissionable uranium is burned and we have a "breeder."

Following mining, the next step in the uranium fuel cycle is isotopic enrichment or concentration of the fissionable U-235 isotope. Environmental impacts from enrichment are basically those associated with the power plants that supply the electricity. However, the presently used gaseous diffusion enrichment facilities (mostly constructed in the 1950s) use large amounts of electricity and cooling water. The equivalent of about 5% of the uranium's ultimately usable energy is consumed in the enrichment processes. Two new processes (gas centrifuge and laser) are being evaluated which use only about 10% of the electricity required for isotopic separation by gaseous diffusion.

Nuclear Power Plants

Where nuclear energy is converted to electricity, a series of barriers separate the process from the environment. These include the steel cladding of the fuel, the pressure vessels surrounding the reactor core, and the massive shell that forms the outside wall of the plant. Elaborate safety systems have been devised to ensure that if anything goes wrong the intense radioactivity inside the fuel rods will not escape. Critics point out that some of the most important features of the safety system have never been tested at full scale and that the theoretical models which serve as the basis for design are not as good as they need to be. In particular, the system to cool down the reactor core in an emergency is viewed by some as insufficient to handle some kinds of conceivable accidents. Presumably, this is not an intractable problem. Given time and research the doubts can probably be satisfied. Meanwhile, the question reduces to: How safe is safe enough? We know that few things can be done without some risk, but there are also risks judged unacceptable by society. In the nuclear question, the issue of alternative risks is very important, even though poorly understood.

Routine emissions of radioactivity from nuclear power plants are only a few percent of the maximum limits established by international health authorities. These emissions can be reduced further but benefit/cost analysis indicates that it would be more productive to use the investment in other ways. Since nuclear power plants tend to be large, and since they don't put any waste heat up the chimney (like a fossil plant does), great quantities of waste heat must be dissipated. For every calorie of electricity generated, nearly 2 calories of heat must be discarded from conventional nuclear plants. Advanced design reactors (the breeder, for example) will only have to discharge 1.5 calories for every calorie converted. We can't do much better than that, for the most optimistic technical experts place the upper limit of efficiency at 1/2 to 1 calorie discarded for every calorie converted, whether the process is nuclear or otherwise; achieving this limit will require the development of new major technologies.

Nuclear reactors are considered by most experts to be virtually failsafe against major accidents. Rather than attempt to try to reduce further the extremely small chance of a major accident, it might make better sense to locate these plants in such a way that should an accident occur the damage will be minimized. There is an analogy in coal-fired plants: rather than try to reduce harmful emissions to zero, the tendency is to locate them away from population centers. In the case of nuclear plants it seems sensible to try to locate them so that the protective buffer area around them can be used for recreation and other low population density uses.

Nuclear Fusion

Once a secret Atomic Energy Commission project with the code name "Project Sherwood," the search for ways to control thermonuclear fusion has been elusive. Some claim the name "Sherwood" wasn't derived from Robin Hood of Sherwood Forest (robbing rich oceans of their deuterium to give energy to the poor), but was chosen because it "sure-would" be nice if it worked! One can find enthusiasts who prophesy the second coming of fusion (its first arrival on earth was a thermonuclear bomb) but most of them are either unbridled technical optimists or their disciples.

It would certainly be wonderful if we could harness fusion for power production. Like nuclear and solar energy, it would not create extra carbon dioxide or alter the world's climate. Also, the problems of catastrophic accidents, radioactive waste management, and terrorist thefts of bomb material would be considerably smaller than with nuclear fission plants. If something were to go wrong with a big fusion reactor (if they work at all, they'll be big), the reaction would immediately cease. There would be radioactivity, but related problems would be minor compared with those of a fission reactor. Unlike coal, fusion would present few routine environmental problems because the fuel is extracted from water and routine effluents would be little more than harmless helium and a few rather short-lived radioisotopes.

There remains one major problem about controlled thermonuclear fusion for making electricity: We don't know how to do it. Eventually, we hope, continued research and development will pay off, but we cannot at present state all major problems, much less provide solutions. Obviously, at best it will be decades before we can count on fusion to supply energy.

ENERGY FROM SOLID WASTES

In our present "effluent" society, we generate between 3.5 and 4 pounds of municipal solid waste per person per day. About two-thirds of this waste is burnable. The rest (iron, aluminum, glass, and the like) contains an important amount of inherent energy due to the energy investment in processing the material from ores. Little wonder, then, when energy prices suddenly increased, that a fresh look at energy/resource recovery became very desirable. There are two main strategies: (1) reduce the amount of materials at the source and prevent their entrance into the waste stream (for example, returnable beverage containers); (2) recapture energy and materials from the waste stream (mine the "urban ore").

Studies of the energy implications of a nationwide switch to returnable beverage containers conclude that consumers would get more for their money and that net energy savings would accrue at the equivalent of about 70,000 barrels of oil per day.

The recoverable energy investment from noncombustibles in municipal solid waste (ferrous and nonferrous metals) is about 4×10^{14} Btu/yr, compared with about 8.5×10^{14} Btu/yr directly recoverable from refuse-derived fuel. The total recoverable energy (using only wastes from urban areas) is equivalent to about 2% of U.S. energy consumption.

Refuse-derived fuel (RDF) contains much less sulfur than oil or coal and can supplement coal in a power plant. The RDF can also be burned in multipurpose boilers to produce electricity and deliver steam for industrial processing and heating and cooling of buildings. By combining both electricity and steam generation, the overall efficiency of the plant can considerably exceed that of the most advanced type of electric power plant. A third use of RDF is its conversion to methane gas or alcohol. Both of these products have numerous uses, basically as substitutes for natural gas and petroleum distillates such as kerosene and gasoline.

Environmental impacts of a major program of resource and energy recovery from municipal solid wastes would be most impressive. If combined with a vigorous "source reduction" program, large numbers of open dumps and other blights could be eliminated while simultaneously easing the demands for new materials and energy supplies. The potential nationwide energy savings alone amount to the equivalent of almost a million barrels of oil per day, even when we restrict ourselves to waste recovery only from the more urbanized areas.

The technology for collecting, transporting, and processing wastes is rapidly advancing but we do not yet have a lot of experience in operating big systems. This is a disadvantage that can only be overcome with time. Perhaps an even greater difficulty is the problem of making the long-term arrangements between various institutions (e.g., industry, municipalities) necessary to organize and carry out such a major systems project. For example, some municipalities will require a charter change in order to make a long-term contractual obligation.

The energy price shift has probably improved the changes of integrated solid waste management far more than any other event during the past two decades. While there is good reason to expect a lot of improvement in resource recovery in the years ahead, we should be careful not to omit attention to the more desirable reduction of materials going into the waste stream ("source reduction"). Many people are inclined to deemphasize source reduction because it can only be done through behavioral changes, however small. At the same time, as consumers begin to realize the full impacts of the "throwaway economy," both on their pocketbooks and on the environment, the impressive potential for energy savings through waste recovery will become an actuality.

SOLAR ENERGY

We already heavily utilize solar energy. It waters our land, cleanses and air conditions our air, and produces our food. "Fossil" sunlight is used as gas, oil, and coal. "Recent" sunlight (wood, falling water, wind) used to be our energy mainstay. Cow dung still is an energy staple in rural India. Hydropower still drives millwheels and electricity-generating turbines. Windmills still pump water from the ground and sails and ocean currents propel many a ship.

In the "developed" nations we gave up windmills, solar water heaters, and other solar devices several decades ago when gas and electricity became so cheap. Now the situation has changed.

The big energy price increase made the biggest difference, but so have (1) our concern to become less dependent on foreign sources of energy, (2) our determination to switch to more renewable and less environmentally threatening energy supplies, and (3) recent technological advances which promise more effective and less costly devices for solar energy utilization.

There are several ways that solar energy can be more extensively utilized in addition to the natural uses we already enjoy.

Solar Heating and Cooling

Thermodynamically, heating and cooling are perhaps the most sensible way to use solar energy. Sunlight is a "dilute" energy source: A bright sun shining directly delivers about the energy equivalence of a 100-watt bulb per square foot of surface. In the United States, the overall average solar radiation is about 17 W/ft^2. Although it is difficult to get much work out of such a low density source it is sufficient to heat and (perhaps) cool buildings because high temperatures are not required. The heating and cooling of buildings and the provision of domestic hot water currently account for about 25% of our entire national energy consumption. Installation of solar heating and cooling systems on 1% of U.S. buildings would save the equivalent of 30 million barrels of oil each year. Solar water heaters, in vogue in southern states during the 1930s, are now a viable alternative in many parts of the country. By using electric heat pumps in conjunction with large water storage tanks, we can even transfer solar energy (or the lack of it) from winter to summer and vice versa.

Environmental impacts of solar energy utilization are not serious; neither are they inconsequential. Well-designed solar heated residences at middle U.S. latitudes need to dedicate most of their southward-facing roof to heat collector surface. A typical U.S. home consumes fossil fuel for space and water heating in winter at a rate of about 8 kW. The average winter solar power incident on a 1000-ft^2 collector would be about 14 kW. Assuming, optimistically, 40% collector efficiency, the collector could furnish about 70% of required heat.

Energy storage or back-up alternate energy sources need to be provided in case of extended cloudy weather. This increases the already large capital costs. Over a six-month heating season a 40% efficient solar collector can save, for each square foot of surface, about 3 quarts of heating oil. Such a system would break even, assuming a 20-year, 9% mortgage, if the solar system, including storage, cost about $\$3/ft^2$ installed. Unfortunately, we're still far from those costs. A variation, however, looks very promising--the use of a massive south-facing wall and associated floor which can capture and store heat in winter but screen the summer sun. This point underscores the essential importance of integrating house design and insulation with solar design.

Building location and design, as well as landscaping, are all constrained under solar utilization. Extra amounts of construction materials such as fiberglass, aluminum, plastic, and glass are required; the provision of these has obvious environmental implications. For example, the production of a pound of alumina requires nearly 10 kWh of electricity, in addition to aluminum and various chemicals. Nevertheless, the environmental impact of a well-designed solar building is attractively small compared with almost any alternative. This fact, plus the fact that sunlight is rather dependable and is "free," underscores its appeal even though there are many practical problems to be overcome if we hope to use solar energy economically and extensively.

Mechanical Energy

Windmills once were a major factor in pumping water from wells and generating electricity for farming and household purposes. Up to 1950, about 50,000 small windmills were in use, mostly in the Midwest. Windmills could become economical again. Environmental impacts of windmills are small except for their visual impact, interference with nearby television reception, and, more important, the environmental pollution associated with manufacturing the windmills themselves. If the present generating capacity were provided by 200-ft-diameter windmills, an array of about one million units would be required, which, if placed side by side, would stretch for 40,000 miles. This number of windmills is about the same as the number of existing electrical transmission towers. We shall return to windmills in the discussion of electricity production.

Energy Production from Biomass

Organic materials produced through photosynthesis can be used to convert sunlight into energy stored as methane or alcohol at an overall efficiency of about 0.5%. Maximum photosynthesis efficiency is about 5% and this value is nearly achieved in tropical swamps, which are about 4% efficient. The average annual values for U.S. agricultural products range from 0.25 to 0.75%. Conversion of organic wastes to methane using anaerobic bacteria can be carried out with approximately 50% efficiency. Enough wastes exist from crop residues and animal manure to produce methane equivalent to 10 to 20% of current natural gas consumption.

Methane and alcohol can be used for a great variety of purposes, including automobile propulsion. Simple molecules such as alcohol contain only about half the energy per gallon as gasoline. As an alternative to conversion, biomass materials can be burned directly in an electric power plant. Biomass contains very little sulfur and the concept of growing materials to fuel a power plant merits careful consideration. However, we must recall that to produce the energy equivalent of one barrel of oil per day requires the dedication of about 15 acres of sugarcane cropland, assuming we can use both the sucrose and the cellulose.

It is clear that we can produce a great quantity of fuel by committing some of our arable land to that purpose, but it makes better economic and environmental sense to first greatly improve the efficiency with which we use energy. It also turns out that physical methods to harness sunlight (which can be used in desert regions) are at least as efficient as biological methods (which require arable land, water, etc.).

Electricity Production

Windmills

In the United States the great plains and the Atlantic Ocean off the New England coast enjoy more wind than other parts of the country. In the contiguous United States, at heights below 250 feet there is more than 40 times as much energy in wind than in hydropower. Winds also increase with height above ground, but so too does the construction cost of a windmill. At favorable locations, wind speeds a couple of hundred feet above ground average 20 mph, equivalent to about 45 W/ft^2 or about three times the average solar energy falling on the ground. Economic trade-off studies indicate that windmills are best constructed with heights of 100-200 feet although some proposals put them higher than 800 ft. For electricity production the optimum generator size is of the order of 100 kW. The use of windmills for pumping water is more attractive economically.

Because the wind doesn't blow all the time, the average generation rate depends on the site, but in Oklahoma, for example, the average power of the wind is about 18 W/ft^2. Assuming a conversion efficiency of about 40% and a windmill radius of 75 ft, we calculate an energy extraction of about 150 kW. Therefore, in order to generate 1000 MW of power, the output of a typical coal or nuclear power plant, we would require nearly 7000 large interconnected windmills. If about 300,000 multiple windmill clusters were installed in a belt from Texas to North Dakota, we could produce about 190,000 MW, or about half of our present generating capacity. In summary, while there is a lot of (solar) power in the wind, it is not cheaply captured.

Ocean Thermal Gradients

As sun-warmed water moving along the surface leaves the equator and nears the poles, it cools, settles, and then moves along the bottom back toward the equator. At some places in the ocean the difference between surface waters and deep waters 1000 or more feet down directly below can be as great as 35°F. Essentially no vertical mixing occurs naturally between these zones. The temperature difference and the vast quantities of water provide an interesting potential for extracting energy from the system. However, the magnitude of the temperature difference makes it difficult to build a machine that can convert the energy to electricity efficiently.

Two or three percent is the best conversion efficiency obtainable from ocean thermal gradients (compared with more than 50% for an advanced thermal power plant). At such low temperatures we could not use a heat engine with water/steam as its working fluid; instead, we would have to use ammonia or a similar fluid that the warm surface water would boil and (after driving a turbine to generate electricity) later would be condensed by the cold water from the deeps.

Despite the complexity of operating a system in the depths of the ocean, with all the accompanying problems of corrosion, growth of marine life on components, low conversion efficiency, getting the generated energy ashore, etc., the idea holds our interest because of the enormity of the energy source. The Gulf Stream, which slides along Florida's east coast, could potentially provide more than 50 times the entire 1980 electrical demand in the United States.

Why don't we commit ourselves to a crash program to extract energy from ocean thermal gradients? Part of the reason is that we haven't had much experience in that technology; we need time to develop it. Also, we're not sure about either the economics (compared with alternative ways of making power) or the environmental implications. It would seem that removing a relatively very small amount of energy from the Gulf Stream would not be harmful. The present U.S. electrical energy demand theoretically could be met by removing enough thermal energy from the Gulf Stream to cause a temperature drop of only about 0.5°F. We can rest assured, however, that our friends in northern Europe whose climate is so dependent upon that stream will be very wary of any tampering.

Direct and Thermal Conversion

More than a half century ago, it was discovered that sunlight could be directly converted into electric current by using certain materials. The space program gave impetus to efforts to improve solar cells; they power many satellites today. Therefore, why don't we use them on earth? The answer is straightforward: They cost at least 50 times more than alternative sources. We're willing to pay the difference on satellites because we have no acceptable alternatives. There is reason to believe that the cost of solar cells can and will be greatly reduced through a major research and development program; prices are dropping steadily in the face of volume production and advancing technology. But this sort of solar utilization won't help us in the near-term even though it deserves major and continuing research support. Environmentally speaking, the solar cell concept is potentially very attractive. The materials required, such as silicon and aluminum, are very common, and, unlike systems using photosynthesis, fertile land, water, and nutrients are not required.

Thermal conversion of sunlight to electricity also can be done today. It's a matter of cost. As the price of the competition goes up, both solar cells and thermal conversion become more attractive. Several ideas are being researched and evaluated. All employ focusing of sunlight with lenses or mirrors to heat a fluid to very high temperatures (e.g., 800°F) so that its vapor can drive a turbine. One system would use rows of black, heat absorbing pipes, carrying the fluid, inside a transparent cover lying at the focus of a long lens. The surfaces of the transparent cover would be coated with a special film so that sunlight would penetrate but heat radiation from the pipes would be reflected back. Another promising approach, called the "power tower," consists of a field of focusing mirrors which reflect and focus sunlight onto a boiler mounted on a tower in the middle of the field.

Both the direct and thermal-conversion systems share the happy fact that they can be sited without competing with agricultural lands. There is a tough waste heat problem associated with thermal conversion but many would relish having to deal with that as compared with radioactive waste management and air pollution from coal.

Hydroelectricity

About 15% of all electric power generated in the United States is produced in hydroelectric plants. It is expected that nearly 40,000 MW may be added to the present capacity (a little over 60,000 MW). However, almost two-thirds of the added capacity will not produce electricity; it will only play the role of storing energy generated in thermal power plants during periods of low demand so that it can be delivered during periods of peak energy demand.

When a dam was built years ago, its purpose (other than flood control, water conservation, navigation, and recreation) was to extract energy from falling water. Now we have pretty well used up the potential for this kind of hydropower.

Big thermal power plants like to run at a steady rate but the electrical demand varies by a factor of two between day and night. It also is higher during the week than on weekends. A pumped storage facility, by design, implies that the water level will fluctuate daily over very wide levels. Spare electricity is used to pump water uphill to a reservoir. When extra power is needed, the water is allowed to flow back down and generate electricity. About 30% of the electrical energy is lost in the process.

Analysis of alternatives to pumped storage to meet peak demand shows that there are some promising opportunities. First, the electrical system demand can have rather sharp peaks because electrical rates for most customers are not time-dependent. In the absence of time-dependent rates, customers will use power at times most convenient to them. Since power at the time of peak demand is considerably more expensive than power at off-peak times, it makes good sense to adjust rates accordingly. In countries where this is done, the electrical demand does not have such high peaks and there is correspondingly less need for pumped storage or other facilities designed primarily to meet peak loads.

GEOTHERMAL ENERGY

Geothermal "sources" are zones of heat that occur mostly in areas of recent volcanism (within the past 10,000 years or so). In the United States most potential geothermal zones lie in the West. The sources of geothermal heat are intrusions of molten rock (magma) that rise to within 2 to 10 miles of the earth's surface. In some cases the magma is forced upward through faultlines or cracks and comes in contact with water seeping down from above the ground. Then we have not only hot rock but also hot water and steam. There are three types of geothermal zones in addition to strictly dry hot rock zones: dry steam fields, wet steam fields, and hot water fields.

The dry steam field is ideal, but rare. The two best examples are in northern California (the Geysers) and northern Italy. In these fields the steam is of such high pressure and temperature that little actual water is brought to the surface in extracting energy. At the Geysers field, about 400 MW is now produced at a very low cost because the 375°F steam is so pure it can be used directly to drive electricity-generating turbines. The California field is apparently one of the largest in the world and may ultimately provide as much as 5000 MW, enough power for several million people, for many decades before the heat source is depleted.

Some people promote geothermal power as "pollution-free." This unfortunately is not so. A 1000-MW facility using "dry steam" would still require about 100 million tons of water per year, containing 1000 to 10,000 tons of ammonia and boron, one million tons of carbon dioxide, and 10,000 to 100,000 tons of hydrogen sulfide.

Unfortunately, most known geothermal resources associated with water are wet steam or hot water fields. In wet steam fields water is superheated under the pressure of overlying rock. After a well is opened about 20% of the water flashes into usable steam but one is left with great quantities of hot waste water, usually loaded with dissolved chemicals (brine).

Hot water fields can be used for heating and cooling buildings or as energy sources for heat pumps. In the United States hot water sources presently known are not convenient to population centers where such an idea might be put into effect, as it has been done in Reykjavik, Iceland, where 50% of the people live in residences heated by geothermal steam. It is now estimated that between 50,000 and 100,000 MW could be generated economically from geothermal heat in the western United States alone, and 10 times this amount around the world. This means that our potential electrical supply from geothermal energy is at least as large as our hydroelectric resources, but hydroelectric energy is renewable whereas economically promising geothermal energy sources are limited in their lifetime.

TIDAL POWER

Tidal power as an energy squrce is in a class by itself because it is derived from the gravitational potential energy given up to ocean tidal motion as the moon draws closer to the earth. There are certain places in which the shape of the land and the ocean bottom near shore create extremely unusual conditions which favor deriving energy from tides. One of the best known locations is the Bay of Fundy, which separates the two Canadian provinces of New Brunswick and Nova Scotia. Engineers have worked on plans for a project there for over a half-century.

The bay "enjoys" the world's highest tidal rise and fall (53 feet) and could provide an electrical output of about 6000 MW at an estimated cost of about $3 billion, which is less than the capital cost of that much nuclear power. One advantage of such a system is that the "fuel" is basically free and extremely dependable. The idea is to build a dam across the mouth of the bay, let tides run into the bay through gates, and let the water run back out through turbines.

If the idea seems so good, why hasn't it happened? The reason, in part, is that it just wasn't economically interesting until the recent big increase in the price of energy came along. Also, since tides only peak twice each day the facility would have to be very complex, with pumped storage capabilities to enable energy to be stored for use during periods between tides. Another reason is that it is not clear what impacts such a massive engineering project would have on the biological productivity of the bay, especially if large pumped storage facilities had to be built. It obviously would also interfere with navigation. In summary, tidal power looks interesting but there are very limited potential sites and actually not very much energy available compared with needs. A few sites may be developed--perhaps in France, the U.S.S.R., or Canada--but for the United States tidal power offers no significant promise when compared with the alternatives.

SUGGESTED READING

A. P. Carter, ed., *Energy and the Environment: A Structural Analysis*, University Press of New England, Hanover, N.H., 1976.

S. D. Freeman, *Energy: The New Era*, Walker, New York, 1974.

J. M. Hollander, ed., *Annual Review of Energy*, Annual Reviews, Inc., Palo Alto, Calif., Vol. 1, 1976; Vol. 2, 1977; Vol. 3, 1978.

A. B. Lovins, *Soft Energy Paths: Toward a Durable Peace*, Ballinger, Cambridge, Mass., 1977.

J. C. Sawhill, ed., *Energy Conservation and Public Policy*, Prentice-Hall, Englewood Cliffs, N.J., 1979.

R. Stobaugh and D. Yergin, eds., *Energy Future: A Report of the Energy Project of the Harvard Business School*, Random House, New York, 1979.

"n" indicates a footnote; "b" indicates a supplemental reading

R

S

T